This remarkable book showcases Ki integrate, summarize and structure a vast range of topics from deeply spiritual teachings to the latest discoveries in science. Kiara brilliantly and eloquently connects the dots, and presents clear and practical solutions for these complex and precarious times. This book will gift you with insights to the past, illuminations in the present and hopes for the future, and will also empower and encourage you to become an active participant in co-creating the New Earth.

Yves Nager, author of Find your Life Purpose

This is a must read for anyone interested in becoming more knowledgeable about the great shift of the ages and planetary awakening that we are in the midst of. I can truly say I have never read a book that is more comprehensive and multi-dimensional. It addresses all levels of this shift, from deeply visionary to supremely practical, from magnetic field collapse and social reconstruction to biological transformation and ascension.

Barry Martin Snyder, co-author of The Luminous Self

As editor of Gaia Luminous, I found its content quite illuminating! The beauty and heartfelt accuracy from which Kiara expresses the true condition of humanity and our planet has superseded any expectations I have had from any editing project. Over my many years as a seeker, following a multitude of teachings, I have never come to experience one that expressed the practical wisdom that Kiara seems to be the steward of. I could read his words again and again, to continually discover new threads of information unraveling, revealing and awakening.

Heidi Mason, "Evolutionary Editing"

Gaia Luminous is a thrilling read. Kiara Windrider serves up a full course of scientific inquiry, evolutionary thinking and sacred insights from a wide variety of traditions. Future scenarios are explored and then woven into the variable of how our collective consciousness may decisively shape the outcome for life on Earth. What emerges feeds our sense of the possible and activates our commitment to a bold new vision for humanity.

James O'Dea, former President of IONS, activist, author, mystic

Gaia Luminous calls each of us to fundamentally alter our human software towards the creation of a better world. This work helps us on our planetary journey into higher consciousness.

Aris Promos Promopoulos, healer

Deep in the human collective psyche is a knowing that the gathering "End Times" chaos heralds a radical transformation for our race and planet. For those who are consciously aware of the challenging implications and seek both cosmic perspectives and spiritual guidance, Gaia Luminous is an invaluable resource of scholarly and balanced investigation, combined with profound personal experience.

Simon Peter Fuller, Founder, Wholistic World Vision.

Gaia Luminous helped me through a very dark patch. Kiara's book is masterly in every sense. Combining extraordinary spirituality with a vast array of scientific research, it points to a supramental vision of our cosmic future while acknowledging the horror and corruption holding sway in our present moments. Your thoughts, words and deeds will be shifted forever after reading this.

Margaret More, evolutionary catalyst

Gaia Luminous is a revelation. The first part could pull many of us into fear, but when read with your inner eyes it is a wake-up call, an invitation to understand the big picture. The second part tells a story of hope and provides guidance on how we can take responsibility for our own awakening and the awakening of the planet.
Setara Dolores Piscador, shamanic healer

Your words touched my heart deeply. The book is to be recommended for everybody who is interested in our planet and the well being of our Mother Gaia.

Heidi Verbeke, owner of "Dolphin Heart"

Kiara has helped me find the tools to understand the painful abuse inflicted by human beings on Mother Earth as well as on animals. Thanks to this book I have been able to simply widen my perception of reality, so I can begin to experience hope once again, and take my place as a true activist on this Earth.

Aldoina Filangieri, animal activist

GAIA LUMINOUS

EMERGENCE OF THE NEW EARTH

BY

KIARA WINDRIDER

Published by:

Kima Global Publishers,
50, Clovelly Road,
Clovelly,
South Africa
7975

© Kiara Windrider 2017

ISBN 978-1-928234-20-3
eISBN 978-1-928234-21-0

Publisher's Web Site www.kimabooks.com
Author's web site www.kiarawindrider.net

Cover design by: Katja Cloud / Cloud 7 Design

Images: Detailed colorful Earth © Janez Volmajer / Fotolia

Stars © Lev / Fotolia

Butterfly: © Michael Stifter / Fotolia

TABLE OF CONTENTS

FOREWORD

By Barry Martin Snyder

Have you found yourself wondering lately about the big picture of planetary transformation? It has been a topic largely on the back burner for many of us, even while so much is being written about the many pressing environmental, social, political and humanitarian issues facing us today. Over the last few years, very little discussion has occurred about the evolutionary process that is unfolding on the Earth as we pass through one of the greatest transitions in planetary history.

Maybe we all became a bit disinterested after all the seemingly failed expectations at the turn of the millennium and again in 2012. Sometimes it feels like we have focused more on the 'trees' of our current social, political and environmental crises and have forgotten to step back and look at the 'forest' of the larger evolutionary process involving the planet, galaxy and multidimensional universe we live in.

Yet, even though we have not experienced the massive Earth changes or mass ascensions that some had expected, we do know that something profound is happening. Accelerating climate changes, global economic crises, a rapidly declining magnetic field, increasing volcanism and earthquake activity, not to mention the myriad social and political crises; all are signs that we are far beyond the point of no return. The idea that humanity may be on a rapid extinction path is now being widely discussed.

Are we going down in flames as a failed species? This is perhaps the biggest question facing humanity today. Outward appearances seem to indicate that this could be so. But on the inner levels, we feel the intensity of deep transformation taking place, and the rate at which we are making consciousness shifts is accelerating exponentially. Awakenings are happening more rapidly than we have ever before experienced. How do these sometimes mutually exclusive dynamics integrate into a coherent picture?

What is going on in these crucial times? What lies ahead? I sense that many are being called to re-visit this whole subject anew. Kiara's new book, Gaia Luminous is a very powerful, direct journey through what is transpiring, and explores the many terrestrial and cosmic factors working to cause the most immense evolutionary shift this planet has ever experienced. It is the perfect book for those who are ready to deepen and expand their understandings.

I have known Kiara for more than two decades. He has become a very close friend, one with whom I have spent many hours discussing the subject of personal and planetary evolution. I have always had a great deal of respect for Kiara's knowledge in this area, which has increased greatly since he sent me a copy of his new book, Gaia Luminous, a revised updated version of Year Zero.

I had spent the last nearly four decades attempting to peer into the great changes that were forecast for these times through both inner and outer exploration. For the last few years I had not managed to find any new perspectives or information on this, and my interest in this theme had waned... until I started reading Kiara's book.

Reading it was akin to putting the capstone on a pyramid of knowledge about what is occurring and our individual and collective journey through it. As I read Gaia Luminous a somewhat latent passion and fire for fulfilling my purpose was rekindled by remembering the immensity and divine perfection of this mega evolutionary event we came here to be part of. The scope and breadth with which Kiara addresses the planetary dissolution and re-birthing process engaged and illumined me.

This book is unique when compared to many of the books of this genre, in that it deals with the most threatening physical challenges directly, laying out the very real issues in a comprehensive, factual way. Rapid climate changes, galactic and solar cycles, volcanism, ice ages, and magnetic pole reversals are all touched upon in a concise and rich way, providing the crucial information that most readers really want and need to know.

Gaia Luminous also journeys through the highest spiritual dimensions of what is transpiring, exploring subjects such as galactic superwaves, global initiation, and supramental transformation, as well as the birth of a new planetary and human species. Kiara goes to the very heights and the very depths of what this great change is all about, and does so without falling into the traps of gloom and doom, nor does he go into avoidance from the real and pressing physical realities by dissociating into an ungrounded perspective, often called spiritual bypassing.

After reading Gaia Luminous I had a much clearer understanding of the many facets of the planetary shift process. They are revealed as interwoven threads, essential aspects of the multi-dimensional tapestry representing the Shift of the Ages. As my understanding has increased so has my fascination and awe for what is unfolding. In the end, I find myself filled with a knowing that Gaia Luminous and Homo Luminous are the ultimate and assured destination, despite any appearances that may seem to deny this reality.

This is the most comprehensive book on the Shift that I have ever read. The physical realities, the cosmic perspective and the dynamics involved as we rapidly approach the shift point or quantum moment, are all addressed. He also includes practices for awakening to assist in making the shift. I invite you to explore Gaia Luminous and be stimulated and catalyzed into greater dimensions of awareness during these times of massive transformation and awakening.

PREFACE

In these times of planetary chaos and upheaval, it is becoming increasingly important to discover the big picture for further human evolution. As Einstein once said, "We cannot hope to find solutions to the crises facing us from the same level of consciousness that created them." It seems clear that if we do not find adequate solutions we will extinguish ourselves as a species, and perhaps as a planet. So where do we look for solutions, and for hope?

With a view towards discovering systemic solutions I would like to offer certain basic axioms, derived from a convergence between scientific and shamanic world-views.

- The Earth is a living system, as is the entire universe

- This living Earth, sometimes known as Gaia, has an evolutionary destiny

- Evolution moves in predictable cycles

- We are currently at a turning point between two cycles

- This turning point could be marked by Earth changes, climate changes, and catastrophic potential

- Humans have an evolutionary destiny within Gaia

- We can choose to pass gracefully through times of planetary chaos and fulfill our evolutionary destiny now

Once our vision and perspective is wide enough, new choices can be made. If we see ourselves on an evolutionary journey greater than our petty struggles to manipulate structures of economic and political power, and greater than our instinctive fears of change, we discover within us the impulse to join together in a grand adventure of consciousness.

And that's what this book is about. Drawing from astrophysics, systems theory, ancient history, metaphysics,

13

geology, shamanic wisdom, evolutionary biology, quantum theory, plasma cosmology, yogic philosophy and futuristic studies, it is an attempt to look beyond the visible domain and to the realm of what is now emerging upon the Earth and within the human species.

This big picture synthesis allows us to face the issues of these times with wisdom and with strength, letting go of fear and denial in the awareness of a deeper evolutionary program that is rapidly unfolding before our very eyes and which we are invited to participate in if we so choose.

Some chapters in this book appear also in an earlier work, Year Zero, although these have been thoroughly revised and updated here. If you would like a free audiobook reflecting key concepts within this book, please go to Kiarawindrider.net. Please subscribe to my list for updates, workshops and travel programs.[1]

PART I

COSMIC PERSPECTIVES

Where God and nature and mankind meet
There I journey with tired feet
To create a world as best I can
Where man lives with nature and God lives with man
First part of a childhood poem, age 8

CHAPTER 1
THE BIG PICTURE

For long eons we have struggled with ourselves and the world, creating veils to foster our separateness, constructing belief systems to hide our ignorance, creating enemies to disguise our inner conflicts, chasing from one experience to another to cover up our inner emptiness, unaware that the same masks that hide our deepest pain also hide our greatest potential.

We are in the midst of a global crisis that is simultaneously peaking on many fronts. We face climate changes, economic crisis, political chaos and environmental catastrophes of unimaginable proportions. It seems clear to me that if something does not radically change on our planet soon, we are doomed to go the way of the dinosaurs. Whether we go out with a big bang in some sort of cataclysmic event, or go out with a whimper slowly suffocating in our own wastes, may not make much of a difference.

Where do we look for hope in all this? Is it possible we can still redeem ourselves as a species? Can we truly learn to live in peace with each other and with our planet? Can we rise to the challenge of these times sufficiently that we can fulfill the destiny we were created for?

"Man is a transitional species," said the great mystic and revolutionary Sri Aurobindo, *"and our highest destiny is still to come."*

Many calendar systems, including the Mayan and Vedic calendars, chart the transition from a reality based in linear time to a reality based in multi-dimensional consciousness. Meanwhile, modern biologists wonder about the possibilities for accessing the unused 97% of our DNA codes, and physicists are commenting on the vast amounts of cosmic rays and gamma rays entering our Solar System, capable of transforming the heart of matter itself.

Is it possible that a vast evolutionary program is unfolding in the midst of human history? Just as the butterfly emerges naturally from a slumbering caterpillar in the course of time, is it possible that we could be likewise emerging into a new biological species, with capacities and capabilities far beyond what we can even begin to imagine today?

Evolution seems to unfold gradually over a long period of time, followed by sharp and sudden changes in genetic patterns. The journey into what many are referring to as the Golden Age or Aquarian Age may not be as long or difficult as we may imagine. We have been preparing for eons for such a change. Is it possible that we may see this happen within the next generation or two?

Astrophysicist Paul LaViolette speaks of a galactic super-wave periodically emanating from the center of our galaxy and moving rapidly through regions of space. This superwave brings with it a new range of cosmic frequencies capable of generating huge shifts in consciousness, shifts in the consciousness of matter itself. Is this what mystics and indigenous people in every age and culture have been looking toward?

I have come to believe that what we term enlightenment in various spiritual traditions is an act of biological transformation within the consciousness of matter. It is not a spiritual event so much as a biological event. It often happens spontaneously for people who have no spiritual background and who are not even looking for it. If this is so, then is it possible that as we go through changes in the consciousness of matter on a collective level, that a wave of enlightenment would follow?

Sri Aurobindo and his spiritual partner Mirra Alfassa, also known simply as the Mother, believed that this would be the case. A hundred years ago, they spoke of the supramental transformation that would be occurring on Earth and emphasized the necessity of experiencing this first in our own bodies. They spoke of waves of cosmic light that would move through human consciousness, preparing us for this great shift.

These waves are coming in now. Many systems are emerging all over the world with the intention of harnessing these waves and using them for inner transformation. My sense is that we are close to a critical mass, at which point the vast reservoirs of undiscovered potential within our DNA will get turned on. We can then harness the incoming cosmic energies and anchor them in the physical body, leading to what some refer to as taking the rainbow body, a process of unifying our physical bodies with our light bodies.

Much of this information is latent in the genetic structure of our physical bodies. I would imagine that a caterpillar has no idea that it will go into a cocoon one day and emerge as a butterfly. We imagine that the limitations we experience within space and time, limitations of disease, aging, and death, are real. And as we believe, within the confines of the linear mind, so it is.

Yet as we connect increasingly to our multi-dimensional consciousness, we are fast entering into a period of history when divine and human are no longer separate, where the body becomes a clear vehicle for spirit, where we can create heaven right here on Earth.

Perhaps this is what the vision of the Mayan calendar, with all its emphasis on galactic cycles, is really about. Perhaps it is what Sri Aurobindo and the Mother dreamed of when they saw waves of supramental light pouring through human consciousness. It is the vision that is imprinted within the biological codes of our own cells, the vision that has been driving us through all these long eons to fulfill our destiny as a new species! It is why we are born, and it is what drives us to remember. It is the dream we have dreamed as divine creators gently awakening now from our ancient slumber.

CHAPTER 2
WAKING UP TO A NEW CENTURY

I will never forget the dawning of this century. It wasn't because of the failed Y2K and doomsday scenarios, which happily, are better forgotten. Neither was it the amazing firework displays synchronized around the world on televisions that weren't supposed to function. It was because of an event that took place that night which showed me beyond doubt that we live in a multidimensional universe.

At the time, I had been living in Mt. Shasta, California, a magical mountain world where the veils between realms are thin. It was early on New Year's Eve. I was on my way to pick up my friend Elisabeth from the nearest airport in Redding — an hour's journey down a gently winding mountain road.

There was a major storm happening. What started off as snow gradually turned into slushy rain as I made my way down the mountain. Visibility was poor. I was tense, not only because of the weather, and not only because I was late, but also because I couldn't help but wonder, in some nagging part of my mind, whether the "Y2K bug" would somehow scramble the foundations of our existing civilization. What might that mean for all of us?

I had filled up the gas tank, and had some warm blankets in the back, just in case....

I was not very focused therefore, and going a little too fast, when I came to a sharp turn around a bridge. The car hydroplaned and I lost control of the steering. I saw myself sliding fast, heading directly towards the edge of the bridge. And then I felt and heard the inevitable explosion of glass shattering and metal rending as I smashed violently against the roadside railing separating me from the deep gorge below.

Although everything was happening much too quickly, some part of my consciousness was experiencing all of it very intently in slow motion. I felt detached from my body, like I was watching it all happen, rather than participating directly. Interestingly, in this state of heightened awareness, there was absolutely no fear, just a curiosity and a knowing that whatever happened, everything was okay.

I can barely explain what happened next. I had seen, heard, and felt myself slam against the edge in that cataclysmic moment of impact. Simultaneously, in that very same moment, I observed with the same curiosity and detachment that the car had come to a stop in a little cul-de-sac on the side of the road, safe from oncoming traffic.

I climbed out to examine the car and there was not a single scratch anywhere. I climbed back in and remained sitting for a long time, shocked, relieved, and stupefied all at once. My body finally started to shake as my mind returned.

What had happened?

Somehow, I realized, I had traveled between worlds, and somewhere I seemed to have made a choice. Inexplicable as it may seem, both events were equally real, and some part of me that was outside and beyond the normal passage of linear time had made a decision to remain on Earth during this time.

Later, as I started drifting back into my 'normal' state of consciousness, the 'other' experience began to take on a dreamlike quality. I found that once I was back in my thinking mind, the cognitive dissonance was too great, and I was forced to choose one reality over the other. From the perspective of the thinking mind, either my body was smashed to pieces in a horrendous car crash or I was safe, alive, and unhurt. The two realities could not co-exist.

Still, for a brief moment I had entered into a mysterious dimension of reality that I could not explain through the laws of classical physics. It was a powerful initiation into the hidden mysteries of time.

This experience seemed a fitting metaphor for our collective journey. We are winding down the path of destiny into new territory. Scientists, mystics, and ancient prophecies alike warn us that we are close to the edge of cataclysmic destruction. We are also told that we are entering the birth canal of a wondrous new creation. What awaits us in these times of simultaneous hope and despair, and how do we choose?

CHAPTER 3
MESSAGE FROM THE FUTURE

Earlier, in August of that same year, a solar eclipse had taken place. Astrologers considered this eclipse particularly significant because it represented a grand cross in fixed signs, signifying entry into a period of global crisis, requiring a deep integration of conflicting or mutually exclusive energies. As we looked towards the new millennium, many of us wondered to what extent humanity, as a species, would be able to make the enormous evolutionary shifts required of us while transitioning through the cataclysmic Earth Changes that had been so widely predicted.

Seeking answers, I went on a vision quest up into the alpine meadows of Mt. Shasta. It was a beautiful time of year, with endless varieties of wild flowers lushly carpeting vast fields of melting snow.

I entered into a period of deep silence. As the days went by, I found myself penetrating beyond the veils imposed by the thinking mind. One morning I woke up feeling a strong presence overshadowing me. It was gentle yet powerful, and felt very, very vast. I felt the urge to write.

Strange words rushed out on paper. The Presence identified itself as Windrider, an 'ascended master and time traveler' who had 'folded through dimensions' from a future point in time to contact me in this moment.

He referred to the fact that we were approaching a great Shift of the Ages, and that there were two streams of probability that stretched out before us as a species.

He spoke first about the nature of time, that it is a function of frequency. Just as there are several notes occupying the same space in a musical chord, there are several tones that we call dimensions occupying the same universe. Each dimension has its own vibrational frequency, its own density, and its own distinct laws of time.

"Therefore," he said, "a being from a higher dimension of the universe experiences time very differently and much more comprehensively than we do. Our own mental consciousness is focalized in a third-dimensional reality where time is experienced as linear — flowing in one direction only from a hazy past into an indeterminate future through a miniscule point of awareness that we call the present."

Higher-dimensional beings occupy the same "space" but with different coordinates of time, or the same "time" but with different coordinates of space. In other words, they hold a multidimensional perspective relative to our own perceptions of space and time in which past, present, and future are all rolled up together into a completely different mode of experience that he referred to as now time.

"In the higher frequencies of density," he continued, "your sense of now time is greatly expanded, collapsing the partitions between past and future and allowing you a much greater experience of the present, and therefore of presence.

"Shamans, mystics, prophets, and meditators through the ages have learned to collapse these partitions and travel to other dimensions, only to return again to what you call the 'real' world. Some, like the Aborigines of Australia, were fluid enough that they knew how to live between the worlds — what they called dreamtime — as their primary reality. In traveling between the worlds, all these early travelers were able to intuit visions of future events as well as access the past through akashic libraries."

He went on to outline a Great Shift that would take place in our future, a dimensional shift in which our bodies and consciousness would undergo a profound change, enabling us to access multi-dimensional realities if we so chose. This is what it means to be an Ascended Master, and this is the state that many individuals will choose to embody after the Shift.

In fact, he said surprisingly, the reason he was able to contact me at this point of time and space was because there was an affinity between us and that, loosely speaking, he was a future aspect of myself from a time beyond the Shift. In the collapsing of partitions that was made possible during my vision quest, the distinction between worlds and dimensions was no longer as rigid.

Our language is not designed for sharing experiences outside of linear time. It is not so easy to describe how it felt to connect with this presence who was separate from myself, yet wasn't; someone who was from the future in linear time, but simultaneously existed in the multidimensional landscape of my own expanded presence.

What was this Great Shift? Windrider continued speaking through the pages of my journal.

"You are at the threshold of an awesome collective event. You are witnessing a cosmic birthing, a zero point long foretold and long awaited, not only by shamans and prophets on Earth, but throughout the infinite galaxies. For reasons that will become clear to you, what is taking place on Earth now is creating a ripple effect through all Creation. A whole new cycle of evolution is beginning... an in-breath towards the Mind of God."

Many of the things he said had been said before, and I had read enough books and talked with enough people that most of this information wasn't new to me. But this time the words were not merely information. They were a transmission that touched me on profoundly cellular levels.

Windrider then launched into the main theme of his message.

"Because of the unique nature of this particular Zero Point, Earth stands at the threshold of a cosmic moment of birth in which Creator and Creation become one. In other words, the veils between the worlds

are about to disappear for a moment in time, or rather a moment out of time, long enough for Creator Consciousness to slip into the hearts and minds of all beings embodied on Earth... it is as if all Creation holds its breath with the Creator, and is then forever after infused with the Spirit of God.

"The frequency of oscillation of any life form has to do with its connectedness with Spirit. In that moment of Infusion, the frequency of all beings on the planet will be raised in one extraordinary jolt of awakening. And because time and space also have to do with frequency, you will be suddenly catapulted into the fourth dimension as a collective event."

By fourth dimension he meant a higher vibrational octave where our bodies were no longer so fixed in space and time, where we could begin to connect directly with the frequencies of our soul.

"While the collective transition to fifth dimension will yet take a while, some of you may find yourself going past the fourth into the fifth or even sixth-dimensional frequencies, activating the ascension process."

The fifth and sixth dimensions involve states of consciousness where bodies of physical matter become increasingly harmonized with what some have referred to as the light body. The ascension process has to do with the unification of the physical body with the light body. Windrider emphasized that many individuals who have prepared themselves for this will be able to achieve this unification during and after this time of Shift.

CHAPTER 4
TWO STREAMS OF POSSIBILITY

But then came the warning. Although he saw from his perspective that this Shift would inevitably take place at some point in the collective future of present-day humanity, how this would happen was still up to us. The future is not fixed in its details, and multiple probabilities can exist simultaneously. Which probability we choose to experience as our reality is up to us.

"The Great Shift I have talked about is inevitable. It is a certainty in all the probable timelines of your future. What actually happens at this Zero Point, or Cosmic Birth, is still undetermined, however, and depends entirely on the collective consciousness latent in that moment."

Windrider described two major streams of probability. The best-case scenario would be that, in the moment of shift, all beings on the planet, and Earth herself, would be translated into a vibrational frequency he referred to as fourth dimension. This would be the starting point for various groupings of souls to continue on even further.

The alternate scenario would be a split between the worlds, where those who were prepared to move into fourth dimension and beyond would do so, passing through the corridors of a cosmic birth. Those not prepared for this would stay in the third dimension, subject to the laws and consequences of our current world condition. More likely than not, they would experience some rather cataclysmic events as the Earth underwent a process of balancing out.

The worst-case scenario, in which the entire planet would be destroyed in a great cataclysm, no longer exists, he further reported. This did seem a likely possibility up until a decade or so ago in our time, but our collective intent and efforts on the inner levels have helped to transcend this.

"A great light is being carried on the Earth right now. This is why we believe it is possible to shift out further from the second scenario of partial translation to the best-case scenario of a full translation into the fourth dimension and beyond."

Christian, Muslim, and Jewish scriptures are replete with references to a day of judgment, where the 'righteous' would be rewarded, and the 'unrighteous' punished. This was the most hopeful scenario available at the time, and represents, according to Windrider, this second stream of probability. Jesus himself spoke of the rapture, of two men working in the fields, one of whom would be taken, the other left behind.

"There are also metaphors in this age that reflect the first scenario. The story of the hundredth monkey, which is an apt description of biological as well as spiritual evolution, is one example. It is always true that when there is a sufficient number of beings envisioning and practicing a new paradigm, the law of grace is then invoked, and the rest of the monkeys, or humanity, as the case may be, are brought into entrainment."

What needs to happen for mass consciousness to shift from the second stream of probability, to the first stream of probability outlined above? Windrider emphasized that the second scenario is based on karma, while the first scenario is based on grace. The laws of grace operate on a higher octave than the laws of karma, but for this to be activated we need to understand and fully embrace our personal as well as collective shadows.

"When the master Jesus spoke of forgiving your enemy he spoke not of condoning their actions, but of recognizing that each of us contains both shadow and light, and that forgiveness provides us with the

opportunity to practice a divine alchemy, unifying shadow and light in the experience of Oneness. He said to love your neighbor as yourself, for indeed you are the same.

"The test for awakened humanity is this," concluded Windrider. "Can you extend the hand of forgiveness and love in this expanded sense to your brothers and sisters in ignorance and darkness — a forgiveness that arises from true knowledge of your own self, a forgiveness that embraces yourself, your partner, your neighbor, your government, the power-mongers, the military-industrial complex, the illuminati, or whatever your own version of the enemy happens to be?

"When you do so, this activates the law of grace, which ends polarized conflict and thereby opens the door to a full planetary awakening. It assures that in the in-breath of God to come, none will be left behind. This is the best-case scenario, and one that I travel into your time to seed as a new possibility."

CHAPTER 5
MYSTERY OF
SCHRÖDINGER'S CAT

The encounter with Windrider was enormously significant for me. Many years ago I read Richard Bach's book *Running from Safety: An Adventure of the Spirit*. The story is based on an encounter with himself in the past. Like all of his books, I was inspired and fascinated, but had never known whether it was fictional or autobiographical.

Now it was becoming clear to me that we are multi-dimensional beings. Once free of the restrictions of the thinking mind, we have access to aspects of ourselves from both future and past, within the eternal vastness of our infinite creator consciousness. In fact, from this perspective, our experience of time itself is no longer the same.

Moreover, what we consider to be our self is no longer limited to a linear progression of a single soul from lifetime to lifetime, but expands further and further out to embrace the entire web of life through all creation in all timelines.

I decided shortly afterwards to adopt Windrider into my own name, in anticipation of what I was growing towards.

This encounter helped me to better understand my New Year's Eve experience. In some way, it prepared me to enter a state of consciousness where I could experience two timelines at once. And perhaps it also provides us with some keys as we look ahead to the complex realities emerging among us today.

Quantum physicists are well aware of strange realities that seem to appear once we begin descending the rabbit hole of subatomic physics. There is a well-known experiment known as Schrödinger's Cat.

Light functions as both particles and waves. In experiments, when light is beamed towards a screen through a double-slit, it creates a wave-like pattern on the other side. While light as a whole creates a predictable interference pattern, each individual photon, or particle of light, can only pass through one of the two slits.

Which slit will it choose? According to classical physics, there should be exactly a 50% probability of taking one pathway or the other.

Schrödinger used this principle to devise a thought experiment. Suppose we take a completely light-resistant box with two slits carved into it. Then we put a pellet of cyanide inside the box, just behind one of the slits. The pellet is designed so that if it is exposed to a photon of light it will release the cyanide.

Next we put a cat inside the box along with the cyanide and bombard it with a single photon of light. If it went through the slit with the cyanide pellet behind it, the pellet would release its poison and the cat would die. If it went through the other slit, the cat would live.

There is exactly a 50% probability that the cat is alive or dead. We don't know if it is alive or dead until we open the box to see. But what if we don't look? What if we don't engage our thinking mind? According to the laws of quantum physics, it is the observing consciousness that makes the choice real, and the cat is in a strange state of suspended animation, neither dead or alive, or both dead and alive, until the observer opens the box to check.

What if the same rules were applied not only to the world of subatomic particles but to the macroscopic world as well? What

31

if multiple realities simultaneously co-exist in our daily experience of life — multiple futures, multiple pasts, and multiple dimensions of the present?

Is this the dreamtime of the Australian Aborigines, or the world of the nagual that Carlos Castaneda referred to? Is there a mechanism that we do not understand as yet whereby this dreamtime world can become our conscious experience, and whereby our conscious intentions can shape the dreamtime worlds? Was my experience on the icy bridge on the eve of the new millennium one example of this? And what of the multitudes of mystical and shamanic experiences people have been having for thousands of years?

If it is the thinking mind that freezes multiple probabilities into one observable reality, are there levels of mind beyond the thinking mind that perceive things differently, which can in turn help to determine which one of these multiple probabilities we choose to download into our immediate experience?

In other words, do perceptions create reality? And what does this mean for us in these times of global crisis?

CHAPTER 6
COSMIC EVOLUTION

Before exploring these questions, let us take a closer look at who we are and where we might be going from a metaphysical perspective. When seen rightly, the metaphysical is not opposed to the physical, and neither is more accurate than the other. They are simply two lenses looking out towards the same reality.

Each step connects with every other step on our journey into forever. In the grand web of cosmic design, each strand is connected with every other, with no beginning and no end. Still, on the linear path of human consciousness there are steps along the way that can be marked. As we journey together, perchance you will find yourself breaking past the cage of linear reasoning into the place where all timelines meet, where all strands of consciousness weave into a single pulse of luminous light.

Humanity is on the threshold of a grand awakening. A cosmic fire inherent within the consciousness of all matter has been pushing us onward for eons, creating long cycles of evolution, moving through sequences of unity and separation, moving through waves of divine descent and human ascent. Born of the first movement of divine light, we have traveled far on the winds of space and time. We await now the moment of return, when all matter becomes illumined with the consciousness of light, and finds its true destiny in the web of creation.

33

This experiment is not new. There have always been the keepers of cosmic fire in the ancient traditions, a few who remembered where we came from, and remembered also how we could return back. Their mission was to preserve this sacred knowledge, and when the time was right, to create a pathway home. If you are reading this, you are likely part of this journey too, and it is time now to remember this sacred mission.

Who are we really? This is a good place to begin. Are we bodies of matter, a product of Darwinian evolution emerging from an unnamed ape in some distant past, evolving into some kind of higher destiny as we journey upwards towards the realm of Spirit? Or are we vehicles of Spirit, playfully creating bodies as we journey down into matter, intent on awakening to our infinite being within the field of matter?

Are we spirit or matter? Our religions teach us one thing, our sciences teach us another. It's a simple question, but what if the question itself was flawed? Our minds are programmed to believe there is a single answer to such a question, that it's either one or the other. Perhaps this program itself is where our problem begins.

Imagine taking a journey into the moment of our origin. What might we encounter? There are so many versions of how life began, who or what God is, endless variations of why we got separated and of how we can return. Which version of this story is true? Does it matter? Perhaps it is time to enter a level of mind where these questions themselves begin to dissolve.

Imagine that there are multiple realities of form, all interlinked and co-existing simultaneously. As humans we experience mental and emotional states, biological realities and, occasionally, spiritual realities. We have a single means for experiencing all these realities, and that is through the medium of what we call the mind.

What is this mind? What if there were not one layer of mind, but many, and it is in the subtle levels of the mind that we experience our identity with God, the universe, and subtle dimensions of light? We are often taught that our attachment to mind is the cause of all suffering. This may be true. Yet we are only talking about one layer of the mind here, the thinking mind, the mind that is caught up in linear time and three-dimensional space.

Is it possible that just as there are subtle layers of the body beyond the physical body so there are also subtle layers of mind beyond the rational mind? Is it within these layers of the mind that we can begin to connect with the corresponding layers of the body, ultimately connecting with a quantum level of physical reality that is well beyond material reality as we know it?

Is it possible that as we connect with the subtler layers of mind and body we then begin to connect with the divinity latent within our humanity? Is this who we are at our highest potential—divine human beings within whose consciousness the veil between spirit and matter has dissolved into oneness?

There are many stories of yogis and masters — human beings who have learned to merge their physical bodies into divine light, and in doing so, transcended limitations of matter, space, and time. Known and unknown, many of these beings still live in earthly dimensions. They work invisibly with the collective morphogenetic field of humanity and await a time when we can all awaken together.

And if this is possible, then how do we achieve it? I feel there is a divine evolutionary impulse actively working behind all the long cycles of human and Earth evolution. This evolutionary impulse works not only in stages of gradual unfolding, but also in quantum leaps of sudden awakening.

This impulse acts not only upon us from forces outside of us in the cosmic field but also from within us. I presume there is an evolutionary timetable encoded within the patterns of our genes which has been gradually unfolding over long eons, a pattern that has been accelerating towards a point wherein a quantum leap of evolutionary consciousness can now happen.

CHAPTER 7
INVOLUTION OF SPIRIT

In order to understand these quantum leaps of evolutionary impulse, we must also understand the force of involution, which could be defined as the process of a cosmic, transcendent, formless, intelligent, all pervading, universal consciousness, creating, entering, and then playing within the field of matter. Evolution and involution go together. We have evolved in bodies of matter, rising upwards through a hidden force within nature, developed organs of life and perception, human beings with a nervous system capable of reaching beyond the limits of mind and reason to touch the great unknown.

And we are also this great unknown, the primal consciousness pervading all things and extending beyond all things, constantly projecting ourselves in and out of physical creation through wormholes of time and space.

We are creation as well as creator. As creation we tend to identify with bodies of matter. We perceive ourselves as human beings with bodies, minds, emotions and senses, tied to a collective past and continually evolving towards new biological possibilities.

As creator, we exist in a realm untouched by time, history, space or matter, unborn and undying, yet externalizing as material creation in the beautiful cosmic dance of Maya.

Our task as humans is to remember both polarities of our being, walk between these two worlds, entering consciously into a vortex where evolution and involution come together as one movement, where creator meets its reflection within creation, and creation wakes up to its potential as creator.

To understand creation we must first contemplate the mechanics of space and time. If the universe started as a big bang, what existed before this? What existed before time itself? And did space exist beyond the cosmic bang? Physicists have long pondered this puzzle and come no closer to understanding this paradox. The laws of physics start falling apart the closer we get to this elusive big bang. Why is it that the equations of quantum reality are so completely differently from the equations of macrocosmic reality?

Scientists such as Nassim Haramein have tried to reconcile these equations by postulating that on a subatomic level each proton is a wormhole connected to every other proton. Other scientists have declared that the only way to harmonize quantum physics with macrocosmic realities is by recognizing that our universe has no beginning and no end, and that all things are held together within a unified field, which the alchemists of old knew as ether.

Later in this book we will also explore the electric universe theory, which resolves this conundrum by declaring that it is not gravitation but electricity which is the prime force in the universe, and that all things are invisibly linked together across space and time through a field known as plasma.

Perhaps all these intuitions are a dark glimpse into the fact that matter does not exist on its own, but is knit within a fabric of primal consciousness that is much more subtle. Within this field of consciousness the substance we identify as matter and energy has always existed and will always exist, visibly or invisibly, even as suns are born and die or universes awake and fall asleep.

We tend to create a dichotomy in our minds in which spirit is eternal and matter is temporal. What if this assumption were false? What if we only experience matter as temporal and limited because our organs of perception are themselves temporal and limited? What if they both exist equally and eternally within consciousness?

Because we tend to identify ourselves with the body, the death of our body in any given incarnation seems to be the end of consciousness. We project this understanding upon the universe, and assume that the material universe also has an end. As our bodies dissolve so the universe must dissolve. As bodies are born so universes are born.

But what if there are not two fixed polarities of matter and spirit but one essential reality appearing as two? What if creator and creation are not two distinct things but one holographic entity within the fabric of consciousness? What if who we are is a single vibrational essence that flows equally through the stars and through the trees and through these human bodies and every subatomic string of quantum existence?

Scientists have shown that the same light that behaves as waves while travelling through interstellar space begins to behave as particles in the presence of an observer. What is it about our mind that creates this observation? Could it be that our human minds are structured in such a way as to literally create an illusion of reality that appears as the world around us, a holographic projector creating a 3D movie?

Our minds are currently structured so that we appear to be separate from each other in terms of space, and move forward in terms of time. But what if we are on an evolutionary pathway where the structure and function of our mind is itself changing to perceive things in terms of unity rather than separation, timelessness rather than time?

Would the laws of physics function differently then? Would we perceive reality in a different way? Would we interact with the world differently? Is it possible that the world out there appears to be what it is only because the collective mind perceives it so? If so, would upgrading the hardware of the mind alter the shape of the world out there?

This, I believe, is the next frontier of human consciousness. As we become conscious of our own consciousness, evolution and involution become a single movement within life's journey. We are making a collective shift from a mental perception of reality based in separation and duality to a supra-mental perception of reality rooted in a unified quantum identity.

Some time ago I began practicing a form of meditation that focuses on flows of sensation and energy pathways within the body. I also began a deep study of the advaita tradition of non-duality, while examining the work of Sri Aurobindo and Mirra Alfassa, two mystics who held that there was a new light 'descending' into the heart of matter, revealing itself within the mind of the cells.

As I gradually began to perceive this truth within my own being, I started to experience states of mind where the skin boundaries dissolved and my experience of self expanded through vibrational resonance to everything around me. It also became possible to enter deeply into any given cell and experience an entire universe within.

The boundary between inner and outer, above and below started to get blurry. I became aware of a subtle vibrational current that connected all things, a current that was the substrate of all matter and creation, but was itself timeless and spaceless. Rather than being an impersonal void, this current was alive with presence, and this presence was the universal self that knows itself at the source of all life.

And this current existed as the spirit within all things regardless of whether I was aware of this or not, whether I categorized it as alive or inert, large or small, visible or invisible. This presence externalized as strings within string theory, or quanta within quantum theory. It projected itself as matter or energy within a framework of time and space, while remaining unchanged and untouched at the heart of all things.

This presence is the evolutionary force latent within matter, while also at the same time the involutionary force that presses down upon the matrix of space and time and spins matter into existence. It is the same intelligence that traverses the galaxies and inspires a lonely wanderer to reflect on the stars at night.

It is the intelligence that orchestrates the planetary web of life, and which nudges us to remember that we are not alone. It is the intelligence that interacts with the higher mind of Gaia to produce a new species of humanity with minds and hearts wide open to this rhythmic dance of creation, arising from eternity and dissolving back into infinity in each moment of existence.

This is the presence I AM, which seems to condense itself in a bodymind complex I have learned to refer to as me. And yes, this body is me, but it is not who I AM, this human personality is me, but it is not who I AM. For I stand alone as I AM, yet have embarked on a journey to incarnate more deeply into this human form than ever before, and to change the very definition of humanness; to enter more deeply into the body of the Earth than ever before, and change the very face of the Earth.

> *The heavens beyond are great and*
> *Wonderful, but greater and more wonderful*
> *Are the heavens within you.*
> *It is these Edens that await the divine worker.*
> *I become what I see in myself.*
> *All that thought suggests to me, I can do;*
> *All that thought reveals in me, I can become.*
> *This should be man's unshakable faith in himself*
> *Because God dwells in him.*

Sri Aurobindo

CHAPTER 8
ENLIGHTENMENT

"This is fullness. That is fullness. From fullness comes greater fullness. Take away fullness from fullness and fullness ever remains."

This ancient affirmation from the Isha Upanishad reflects an understanding of perfection unfolding in time that the thinking mind cannot easily grasp.

The thinking mind is designed for complexity. We compare, judge, discriminate, analyze, interpret, divide and subdivide everything from writing down this simple sentence to attempting to discover the meaning of life and the universe.

The mind is a beautiful tool for learning to embody ourselves in human incarnation, but we have identified with the mind, thinking we are this mind, or that this mind is the lens to discovering who we are. The more we have identified with the mind, the more complex our lives have become, and the less we can be present to the profound simplicity and beauty of each spontaneous moment.

We assume we are a body-mind-personality complex. We are born and then we die. We have a name and we create a history, we possess thoughts, feelings and memories. We think we are real, in the sense that this entity that we call by name is permanent, at least until we die.

We also assume we are separate from the rest of the world, other objects, other people, nature, the universe, God. We struggle to find our place in this complex world, struggle with ourselves and our inner conflicts, struggle for meaning and purpose so we can feel worthy, happy and loved.

We get tired and exhausted from this endless struggle, wondering why it is we have lost our spontaneity and joy. Some of us embark on a spiritual path and undertake a quest for the true self. We read all these fascinating stories of spiritual masters who have achieved a certain state of consciousness where they transcended the limitations of the world around us, a select few who have managed to escape the drab realities of an egoic world view and achieved a transcendent blissful unity with all of life, creation and the universe.

We are told in these stories that the world is not real in itself, that who we think we are is an illusion, and that the only thing that is real is the Self. We are told that we must give up everything in order to realize the Self and that once we do so we are liberated from all our suffering and can dwell in a state of permanent enlightenment. Enlightenment is presented as the ultimate experience, the end of the path, the final goal of all our seeking.

We go desperately seeking after this thing called enlightenment, hoping it will provide us with the spontaneity and connection we have lost. We forget that the thing we are seeking through is the same mind that craves complexity, and that the one we are seeking for has never actually ever been lost.

In fact, **that** which we are searching for is the same Self that we already are, and always have been. This is the great cosmic paradox. We can never find this Self by seeking after it, simply because it is already **that** which exists behind all seeking, all exploring and all being.

Ironically, even our identity as a spiritual seeker is sourced in the duality of the thinking mind, which is forever attempting to create expectations and judgments based on good and bad, right and wrong, spiritual or unspiritual. We think we are somehow flawed and therefore engage in a quest to change ourselves. We discover it is impossible to change the nature of our mind, and end up suffocating in guilt and shame.

We embark on a journey of endless seeking and constant self-analysis, gaining a profound understanding of our attachments, aversions and addictions, while still unable to gain mastery over them. We idealize those we consider more advanced than ourselves, fueling the search, and separating ourselves further from our own beauty and divine glory.

We give our power to those we consider enlightened, feeling inadequate and small in comparison. We put them up on a pedestal because we feel inferior and then send them crashing down in order to feel superior. We are constantly comparing, judging and analyzing each other as well as ourselves. Just like with the rest of our daily lives, the core belief is that we must do something, get somewhere or become someone in order to feel adequate.

So what's the way out?

We must realize that this is all part of our evolutionary design. Human evolution is not yet complete. Constantly comparing, analyzing, justifying, or judging ourselves doesn't help. The thinking mind cannot break out of its limited role by itself and function. Or as Einstein said, we cannot solve the problems we face today from the same level of consciousness that created them. So what then?

Let us examine some of the advaita teachings of ancient India to understand more clearly what this elusive journey is about. Perhaps we will find that this journey of enlightenment is not about going anywhere but recognizing where we already are, not about doing anything but anchoring more deeply into the reality of what is already here.

Discovering the Self is not difficult. It is not the end of a long tortuous path but the very beginning. The Self pervades all things and is at the very root of our awareness, thoughts and feelings. It is the sense of being that exists before thought arises and remains in the midst of every thought

It is the consciousness that animates our body, the consciousness that inspires our thoughts and feelings, the consciousness that provides us with a sense of identity through the individuation of the mind. And it is the same consciousness that fills the entire universe. It is inseparable from our very existence. It is the presence we experience when we simply take a moment to be quiet, nothing more mysterious than that.

Our journey begins with the self-evident truth that this presence is who we are. This first stage of awareness is known as Self-realization. Since this same presence vibrates at the heart of everyone and everything, we are essentially not separate from each other, whether it is another person or a flower, a rock or a galaxy.

When this awareness becomes deeply established within our human experience we refer to this as enlightenment. The teachings are very clear that enlightenment is not a mystical experience to be sought or achieved, but rather a deepening realization based on an accurate understanding of who we are.

This takes time. Our conditioning within the illusion of Maya runs deep, and it takes awareness and effort to break out of this hypnotic web. But eventually, the ego identity diminishes and we become established in our identity as the Self.

The advaita teachings begin with the premise that before the creation of matter and life, the Self has always existed. The Self is not bound to creation, time or space. The qualities associated with the Self are Existence, Consciousness and Endlessness, and since all things that exist share these qualities, all things share a common existence.

Consciousness exists independent of matter but matter cannot exist independent of consciousness. In order for matter to exist, Universal Consciousness, or Brahman, must create a reflection of Itself through the agency of Maya, the principle of projection. This creates the realm of Universal Mind, Ishwara, which expresses itself as the physical universe in the fabric of time and space.

Our essential Self, or Brahman, eternally exists before time, beyond space, and outside any definition of energy or matter. This Self possesses all the qualities we might associate with the consciousness of Creator, Spirit, or God. Like a mirror shattering into tiny pieces, each piece reflecting the whole, so Brahman projects Itself into an infinity of soul-sparks, creating the illusion of separateness and multiplicity within the realm of Matter.

Who we are therefore is the essential Self, projected into the matrix of Creation, and extending into bodies of varying density within the fabric of time and space. The human mind is a holographic reflection of the universal mind. Our sense of egoic identity is a function of this mind, and real only in the sense that a reflection in the pond derives its reality from THAT which is being reflected.

As we become aware of this truth, and learn to carry this awareness into all the circumstances of our lives, we achieve a freedom from identification with the separate ego and recognize ourselves as one with the infinite source of all things. As this recognition grows within our human experience we come into the awareness that nothing happens by chance, and that we are the creators of every aspect of our life. In this knowledge, we set ourselves free.

We understand that there is no fixed reality out there in the world controlling and directing the circumstances of our lives. Rather, it is the structure and conditioning of the mind that generates the illusion of a reality separate and outside of ourselves. Morpheus might have referred to this apparent reality as the matrix.

Once we understand the nature of this matrix, we discover that we can release ourselves from the mass hypnosis that governs our human experience. We can take the red pill, unplug ourselves, and become conscious co-creators of our personal lives and of our world.

Being a co-creator in this context begins with the recognition that we are infinite beings of limitless power. This truth derives not from the limited identity of the body-mind-personality-complex, but rather from an alignment with universal consciousness that is our essential Self.

If we are one with the Self, and the Self pervades all things, we are no longer subject to the fears, illusions and perceptions of the mental ego. We learn that existence is multidimensional and that as we allow this multidimensional force to move through us, we gain access to a synchronistic flow of life, generating ripples of creativity, beauty and joy wherever we go.

We cannot be creator and victim at the same time. As we deepen into our identity as the Self, we no longer experience ourselves as victims of circumstances or subservient to the powers that be. A self-harmonizing force begins to emanate through us that is capable of touching all the people and circumstances of our lives.

Enlightenment then, is not an escape from our own dysfunctional relationships with ourselves and the world. It is about taking responsibility for the divine fullness that we are, going beyond the borders of our mental conditioning and emotional prejudices, and entering into a higher spiral of evolution as global citizens and universal humans.

CHAPTER 9
BEYOND DOING

"We shall not cease from exploration. And the end of all our exploring will be to arrive where we started. And know that place for the first time"

T.S Elliot (Little Gidding)

Many people equate the evolutionary journey of the human species with enlightenment, and I feel it is important at this stage to distinguish between the two. Enlightenment is a state of consciousness in which we are no longer identified with the personal ego. It is a freedom from the restrictions imposed by our conditioned mind. However, the evolutionary journey I am referring to is a collective event, a genetic transformation of the human species. It is a process of planetary evolution initiated by external galactic factors.

Given this distinction, it may be useful here to define enlightenment in the context of our evolutionary journey. Though one of the most mystifying of spiritual concepts, it is important to understand clearly what it means, as well as what it does not mean.

Enlightenment is not so much about becoming superhuman, turning into light, manifesting spiritual powers, or transcending the murkiness of daily life and emotions. It is not about cosmic bliss, instant clairvoyant abilities, or escaping from human

Gaia Luminous

responsibilities. It is not about losing the mind, or even somehow changing the nature of the mind.

There are many layers of the mind relating to many layers of consciousness, as I discuss in detail later in this book. As we have seen in the last chapter, we are conditioned to perceive reality through just one of these layers, the thinking mind. The thinking mind has an essential role to play in our journey of embodiment, but somehow we have become conditioned to believe that this is the only level of mind that exists.

The thinking mind perceives reality through its ability to compare and analyze information. It clothes itself with a set of subconscious beliefs and emotions, which we all inherit from a collective web of human conditioning. It is based on memory, and creates a self-identity originating from a sequence of events in linear time. Referring to this sense of self as the personal ego, we experience ourselves as a fixed identity starting with birth and ending at death.

Enlightenment transcends the duality inherent within the personal ego, leading us to understand and celebrate the natural flow of life, without craving for something that seems desirable or pleasant nor resisting what we consider painful or unpleasant. We are no longer attached to the mechanisms of comparison and judgment inherent within the thinking mind.

Freed of the web of conditioning, we gain the ability to express ourselves spontaneously in each moment, living from our authentic self rather than from conditioned responses. Freed of the burdens of the past or expectations of the future, we learn to engage with life in present time. Freed of our identification with the thinking mind, or personal ego, we begin to incarnate through these body-mind organisms as spontaneous expressions of divine consciousness, or soul.

The question inevitably comes up, *"Okay, all this is fine, but what do I DO in order to get enlightened?"* We must understand that our personal ego is based in doing, and the thinking mind cannot stop the thinking mind. Once we truly understand this, paradoxically, we can then cease our efforts, accept our failure and pain, embrace our inconsistencies and shadows, relax in the truth of what already is, and make peace with the universe. As we do so, we realize that in the very act of letting go we have already found what we have been looking for.

A couple of years before he died, I attended a Satsang with Ramesh Balsekar in Bombay. A close disciple of Nisargadatta Maharaj, Ramesh is considered an enlightened master in his own right. Much of his teaching has to do with letting go the idea of being a doer.

"Our basic problem is the identification with ourselves as the doer," he addressed the group of students surrounding him. *"As long as we are identified with the doer, we think we have choices in life, chasing pleasures, avoiding pain. We think we are making wrong choices and feel guilty. Or we make choices because we are afraid to live out our own truth. But if every action we take is simply a result of our conditioning and our genetic inheritance, then all our choices are based in fear. Do we really have free will?*

"We are free only when we no longer identify with the doer," he continued. *"Then we become a doing, and life becomes a happening. No longer identified with the doer, we no longer live in fear of making the wrong choices, or that somehow the universe can do us harm. We release our experience of guilt, fear, blame, and regret, engaging spontaneously with life in the present moment, knowing that everything that happens is part of God's will, divine destiny, and cosmic plan.*

"The problem is not with the ego," he emphasized. *"We are all subject to the subconscious influences of the personal ego, the psychopath as well as the sage. The difference is, in the case of the enlightened sage, the sense of personal doership has been uprooted.*

"If we are not the doer how can there be karma? Karma is only real for us if we are identified with the physical form. Once beyond this limited identification, we are free to manifest the full power of our divine destiny."

Who am I then if not the doer? Balsekar encourages us to become a witness to divinity passing through us, creating itself in each moment fresh from an infinitely creative source, according to its own wisdom and its own timing. As we practice shifting our identity from the doer to spontaneous doing, from ego to source, we realize that all things happen in a continuous stream of divine will. Our security then comes in knowing that we are not separate from this stream, and can therefore surrender to this flow in absolute trust.

This is a perspective that our limited body-mind organism cannot easily grasp. Enlightenment is the realization that all things in creation, no matter what the appearances, are part of this divine flow, perfect in themselves in the given moment, and unfolding gradually towards an even greater perfection, a cosmic plan.

All we can do is accept and understand the perfection of our place within the whole, exactly the way it is. As soon as we acknowledge this, the curtain lifts. Our obsession with our personal journey or with personal enlightenment ends. We realize that we are not so much an actor on the stage of life, but life's own desire to express itself deeply, lovingly and powerfully in each moment of existence.

CHAPTER 10
ALTERNATE VIEWS OF
TIME

Over the years I have noticed that my quest for personal enlightenment no longer drives me like it used to. Perhaps I have come to recognize that this is a lifelong process rather than some ultimate event. Perhaps I am no longer quite so identified with the personal self. Perhaps I have learned to be a little kinder towards myself and others, more tolerant of shadows and imperfections, and more spontaneously willing to engage the flow of life arising from someplace bigger than the third- dimensional realities perceived by the thinking mind.

What I feel motivating my life instead is a deep passion for a shared planetary awakening, for understanding the timing and mechanism behind this mystical birthing process, and for joining with others in co-creating new worlds in space and time. We are at a pivotal point in human history, and none of us can move forward unless we do it all together.

So where do we go from here? How do we better understand and participate in this process of birthing?

One clue to our evolutionary process may be found within the ancient calendar systems of earlier civilizations. There are Egyptian, Aztec, Hopi, Inka, Cherokee, Hindu, Chinese, Tibetan and various other calendar systems that all share certain common perceptions of evolutionary history and time cycles, although they also differ widely in their details.

Many of us had become quite obsessed with the Mayan calendar some years ago. This calendar is actually the last remnant of a much older calendar system that goes back to the beginning of time, as we understand it, and also pin-points the end of time as we currently experience it. It is not so much a historical calendar as an interface between linear time and what we might call galactic time. As such, it can be interpreted as a map of our evolutionary impulse from linear history into multi-dimensional consciousness.

Consciousness is inseparably linked with our reckoning of time. The ancient Mayans, in attuning their sense of time to a galactic rather than solar calendar, were able to attune themselves to a multi-dimensional rather than linear consciousness.

What does this mean for us today? Our civilization has become increasingly bogged down in a linear framework of time which increasingly separates us from the awareness of multiple, inter-connected realities. Linear time functions in one direction only, leading from a fixed past through an ephemeral present into an indeterminate future. We experience ourselves as a biological clock ticking away the moments between birth and death. We experience history likewise as a linear progression of events in a single dimension of reality.

Galactic time, however, orients us to an expanded world of simultaneous realities, each linked together in a quantum field of infinite possibilities. This view of time opens us up to experience a cosmic consciousness where we become simultaneously aware of multiple dimensions of being, all extending out and inseparable from our divine Self. We are no longer an ego-based personality trying desperately to connect with our souls to give meaning to our puny existence, but rather the vast landscape of divine soul simultaneously experiencing ourselves in time and space within bodies of matter.

Please realize that we are not just talking about a calendar system but of the nature of consciousness itself. Perhaps as we begin to truly experience galactic time we will discover ourselves to be galactic beings, no longer subject to the limitations of linear timelines that have imprisoned us for so long in disconnected boxes of colorless perception.

Our level of human consciousness determines our linear experience of time. Linear time is associated with a perception of reality that is primarily mental. Galactic time however, has to do with higher levels of the mind, which offer multi-dimensional perceptions of reality, multiple time-lines.

As long as our sense of reality is determined by linear time our choices are limited. We become passive observers in world events that often seem chaotic and overwhelming. If this is our only window into reality it becomes very easy to fall into a black hole of powerlessness, hopelessness and despair when confronted with the insecurities and chaos all around us.

As we begin to connect with galactic time, however, we recognize that we are creators of reality to a degree we can hardly begin to imagine. As quantum physics is beginning to understand, our experience of reality is totally reflected in our beliefs about reality, and external realities can change instantly in response to an expansion of perceptions.

This is a science we are just beginning to understand. Physicist Michio Kaku, in his books, *Physics of the Impossible*, and *Physics of the Future*, lays a groundwork for the possibility that multiple timelines co-exist in consciousness, and that we can choose to make real whichever timeline we focus on.

Could it be, as I experienced during that mysterious episode on the bridge at the eve of this new millennium, that we have already averted many potential catastrophes by the power of our unified consciousness? Could we literally create a paradise on Earth based on our understanding of these higher principles? Are we approaching a long awaited moment when Creator consciousness slips through the veils of Creation, when we begin our journey into multi-dimensional consciousness, when the partition between the worlds begins to dissolve?

This is a vision that inspires me, and so many others, during these turbulent times.

PART II

PHYSICAL REALITIES

Where is the thicket? Gone. Where is the e agle? Gone. The end of living and the beginning of survival...
Chief Seattle of the Squamish, 1854

A human being is a part of a whole, called by us 'Universe', a part limited in time and space. He experiences himself, his thoughts and feelings, as something separated from the rest — a kind of optical delusion of his consciousness. This delusion is a kind of prison for us, restricting us to our personal desires and to affection for a few persons nearest to us. Our task must be to free ourselves from this prison by widening our circle of compassion to embrace all living creatures and the whole of nature in its beauty.
Albert Einstein

CHAPTER 11
ENDANGERED PLANET

I spent my high school years at an international school in the hills of South India. On weekends, a small group of us would often hike the breathtakingly beautiful surrounding forests. We would go for days without meeting another human being, and would often encounter herds of mountain goats, wild elephants, and bison. Occasionally, we'd even spot the evidence of tigers in the vicinity. The hills were covered with shola, indigenous jungle vines and vegetation. The streams sparkled with the magnificent shimmering of rainbow trout.

Little of it now remains. The last time I visited, the forests were gone, the streams were dry, and the beautiful mountain village of Kodaikanal had become choked with auto fumes, cheap tourist hotels, and mounds of empty plastic bottles. I have not been back for many years now because I wish to remember this mountain paradise the way it used to be.

What are we leaving behind for future generations? Although there are millions of us who genuinely care, our species as a whole has become rather callous towards the fate of our planet. Mimicking cancer cells, we have embarked on the accelerated suicidal course of destroying our own habitat. The forces of greed, ignorance, and denial, alongside our belief in short-term profits at any cost, coupled with the politics of human survival, seem much too pervasive for us to willingly step towards change extensive enough to heal us.

I will cite some facts and statistics in the next few chapters, putting them together in new ways so that we begin to realize how vital these issues are, how directly we are impacted, and how little time we have left, if any, to turn things around.

Much of this material, if you haven't heard it before, might cause you to spiral deeper into helplessness, guilt, or fear. That is not my purpose here. Although it is important to be aware of these things, guilt, fear, and helplessness only serve to immobilize us further. Please don't get stuck there. Eventually I will attempt to broaden our perspective to look at the evolutionary big picture, which might provide us with some clues and even some solutions, although not in the way we might expect.

As I continue to research global cycles and events, there seem to be increasing warnings of current and imminent Earth Changes, usually defined as dramatic changes in climate, accompanied by geological events such as earthquakes, volcanoes, hurricanes, droughts, and flooding. Some refer to events of even greater catastrophic magnitude such as magnetic reversals, super-volcanoes, crustal displacement, meteor impacts, and pole shifts.

What do we do with all of this? Some of it falls into the category of disinformation or hoax, deliberate attempts to mislead or terrorize people in order to control them. Some of it is saturated with elements of our own subconscious fears, often based on racial memories of earlier catastrophic events. Much of it, however, is carefully researched information derived from seismologists, climatologists, astrophysicists, geologists, ancient record-keepers, and contemporary seers and mystics.

In this section I would like to examine this information through the eyes of science, based on experimental observations of the physical world. Much of this research may seem somewhat frightening and grim. If you have come this far, please don't stop with this. I will attempt to show in later sections of this book that these same expressions of gloom can be the prelude for an incredible vision of renewal and hope.

It is important to be discerning, for, yes, there is a lot of misinformation out there. However, we are seeing tremendous changes on the Earth today, with more in upcoming years, and I feel we also need to be aware. Awareness of changes is not the same as going into fear. In fact, a certain degree of fear can be a healthy antidote to denial. But we need to acknowledge that we are already in a time of dramatic Earth changes, much of which goes unreported in the mass media.

Many of these Earth changes are caused by human willfulness and interference. We have created a nightmare of environmental destruction, slaughter, and pollution in the name of economic growth and human progress. What we are likely to see in the near future, as a result of disharmonic patterns we are creating today, is terrifying. In a few more decades conditions on Earth could become unfit for almost every form of life.

We cannot heal until we are willing to find the causes and extent of our injuries. The physical body of our Mother Earth has been pushed out of balance. We cannot ignore this. She is a highly evolved living entity. She is attempting to heal herself by correcting some of these imbalances. Some of this healing will involve Earth changes.

Once we understand this, and if there is enough conscious intent, healing can happen. This is the power and potential of the divinely inspired human spirit. But where is our power if we are asleep or in denial? Where is our power if we are still governed by fear?

We need not go into fear. If we look at Earth changes from an evolutionary perspective, what we see is much bigger than the Earth simply trying to find a way to heal Herself. Supported by cosmic beings and a vast galactic network that includes our own higher-dimensional selves, She is involved in birthing an entirely new reality, more expansive than any of us realize.

Our minds are too conditioned by the past to conceive a passage of this magnitude. I see a transformation taking place that is unprecedented in the history of our planet, involving not only our own planet, but also our entire Solar System, galaxy, and beyond.

The conclusions derived here are confirmed by scientists and mystics of the first order. This gives me hope. And more than hope, it gives me perspective. It is not an ostrich perspective from underneath the sand, but a systems perspective that is bigger than our individual selves.

From this perspective I see that the Earth changes to come are not something to fear. These are simply the necessary adjustments we need to make so we can move forward in trust as spiritual beings inflamed with a planetary vision, moving forward beyond the nightmares of our misqualified creations into a divinely inspired heaven on Earth.

I understand that all of us are conditioned by an instinct to survive, and this instinct makes us recoil when we hear of Earth changes, suffering, turmoil, and destruction. Many of us on a spiritual path have also been conditioned by another instinct, an instinct that refuses to acknowledge destruction as an aspect of transformation, or death as an aspect of rebirth.

This is ironic when we consider the teachings of Christ, Krishna, Lao-Tsu, or any of the great initiates who have graced the Earth. It is especially ironic knowing that our civilization has learned to peddle the dark side of death to the extent that governments can get away with every kind of murder and justify it in the name of protecting our 'way of life'.

What does being on a spiritual path truly demand of us? Do we understand the long-term process of change? As greater frequencies of light enter our planetary fields, whatever darkness needs to be cleared is illuminated. We are moving through the tunnel of planetary birth, and birth is sometimes painful. A true spiritual transformation is not a tame, sterilized process in which we can keep a dysfunctional way of life intact. Transformation demands that we 'boldly go where no one has gone before.' As romantic as this may sound on Star Trek, it is the hardest thing on Earth for any of us conditioned by attachments to the past.

Transformation requires opening our minds to new perceptions, our hearts to new sensitivities, and our souls to new possibilities. Human beings are so constructed that this does not happen unless we reach a crisis point. If there are Earth changes to come, it is only Mother Earth trying to shake us awake when every other means has failed.

We have every right to be fearful as we step into the future. Yet, let us not lose sight of what this planetary adventure is all about. If you have ever surfed a big wave, or skied a double diamond, or jumped off a high waterfall, you know the knot of fear in your belly as you leap off the edge, only to dissolve into profound celebration as you sense a taste of life beyond that edge. It is the same with us collectively as we step into the tunnel of planetary and cosmic birth.

CHAPTER 12
GAIA HYPOTHESIS

It is becoming increasingly difficult to deny that our planet is in trouble. Top scientists all over the world have been warning us for years that Earth and humanity are seriously endangered. We are all familiar with terms like nuclear radiation, greenhouse gases, ozone depletion, extinction of species, global warming, pole shifts, space weather, and more recently, global cooling. But what does all this really mean for human and planetary survival?

A patient must know they are sick before they are willing to look for a cure. My intention in the next few chapters is to very briefly spell out the nature and extent of our sickness, and the symptoms we are currently experiencing. Later we will focus on the cure.

To understand our dilemma rightly, we must first look at how we perceive the Earth. Do we see this planet as a dead hunk of matter that we can plunder, pollute, and exploit at will? Can we continue raping and pillaging without consequences? Are there limits to what the Earth can handle? What would happen if she decided to fight back for her own survival?

Indigenous people in every part of the world have always regarded the Earth as a living being, whom they refer to as Mother, or sometimes Grandmother. This high regard was a way of honoring this beautiful home planet, acknowledging that she was a sentient being, and affirming how deeply they felt nurtured, sustained, and protected by her. They also spoke about a delicate web that connected all life on Earth, and even beyond.

Pocahontas, the Walt Disney film based on historical characters, conveys the difference between indigenous attitudes towards nature and the ignorance and insensitivity of most European settlers in the New World.

As Pocahontas, daughter of a tribal chief, tells Captain John Smith:

"You think you own whatever land you land on. The Earth is just a dead thing you can claim. But I know every rock and tree and creature has a life, has a spirit, has a name... Have you ever heard the wolf cry to the blue corn moon, or asked the grinning bobcat why he grins? Can you sing with all the voices of the mountains? Can you paint with all the colors of the wind?"

In personifying Nature, Pocahontas came into relationship with Her. To the original inhabitants of North America, and all indigenous people around the world, this was a way of life. It was a perception of reality that shaped their experience of the world. They could not dream of taking from the Mother without giving back. They took only enough to sustain themselves. To live in harmony with the web of life, was to walk the beauty way, and it was an integral part of their very existence on Earth.

"Man did not weave the web of life," said Chief Seattle of the Squamish people 150 years ago. *"He is merely a strand in it. Whatever he does to the web, he does to himself."*

Chief Seattle has become well known for his impassioned speech when he was asked to 'sell' his land to the European settlers. Although there has been some recent debate about whether these were his exact words, the message still stands. *"How can you buy or sell the Sky, the warmth of the land?"* he pleaded. *"If we do not own the freshness of the air and the sparkle of the water, how can you buy them?"*

All the memories and identity of his people were linked to the relationship they felt with this land. Was it so easy to give this up for beads and blankets? He realized the inevitable truth, however, that his people had been broken by starvation and war, and that they were going to lose their ancestral homeland to the mad rush of European settlers who passed themselves off as the legitimate government of a country that they had stolen.

Before signing the treaty, Chief Seattle offered them the natural wisdom of his people. His speech is beautifully poetic and a haunting reminder of what most of us have forgotten. *"This we know,"* he concluded, *"the Earth does not belong to man. Man belongs to the Earth. This we know. We are all connected, like the blood that unites one family."*

The few surviving indigenous people still understand this, and still speak out wherever their voices can be heard. In recent times this systems perspective has become popularized by British scientist Dr. James Lovelock, who demonstrated a scientific basis for the interconnected web of life, which he called the Gaia hypothesis. Lovelock refers to this living Earth as Gaia, named after the Greek goddess of the Earth.

He was able to put into scientific language what the indigenous people have been saying all along — that the Earth is a living super-organism, with intricate and interconnected feedback systems just like any other organism.

In his first book, *Gaia: A New Look at Life on Earth,* Lovelock outlines how Earth is a self-regulatory organism with dozens of negative feedback loops that keep our planet functioning within stable limits. Just like our bodies have feedback loops that govern our temperature, circulation, respiration, nervous and immune systems, so does the Earth.

For example, our bodies develop a fever when we get sick, which redirects the energy flows of the body, and fries up any invading viruses. Likewise, Gaia has her own feedback systems to help her recover when she gets sick or thrown out of balance.

Let us ask ourselves whether there is a tipping point beyond which she cannot recover, just as the human body cannot survive a fever for too long before it dies. In our senseless exploitation of the natural world, are we rapidly killing off the only home we have? As the title of his latest book, *The Revenge of Gaia,* may suggest, Lovelock himself is not so hopeful that we have either the will or the capacity to pull ourselves out of a rapidly descending spiral. Have humans become an invading virus? Will Gaia eventually incinerate us?

CHAPTER 13
GLOBAL WARMING

In 1992, the Union of Concerned Scientists issued a warning. This included 1,700 scientists, including 99 Nobel laureates, from 69 countries around the world. They emphasized that we are on a collision course with nature, that human activities have inflicted harsh and irreversible damage towards the environment and critical resources, and that fundamental changes are urgently needed if we are to avoid this collision.

The report highlighted several areas of stress: our atmosphere is polluted; our ozone layer is depleted; industrial wastes returning as acid rain are killing our soils; lakes, rivers, oceans are likewise seriously polluted; increased levels of acid in the oceans from toxic chemical runoff is threatening the whole chain of life beginning with sensitive coral reefs and coastal ecosystems; fisheries are rapidly declining; and worldwide water shortages are becoming increasingly severe.

Heavy erosion of topsoil as a result of negligent farming practices, as well as increased chemical toxicity from indiscriminate use of pesticides and fertilizers, are affecting productivity. Meanwhile, heavy deforestation of tropical and temperate forests is likely to dramatically alter weather patterns, contributing to fires, hurricanes, droughts and flooding. It is estimated that most of the rainforests will be gone by 2100.

A third of all species now living will be lost forever by 2100, they further predicted, unless drastic steps are immediately taken to prevent these trends.

The report went on to suggest what we would need to do to reverse the damage. This included moving away from fossil fuels, halting deforestation, conserving and recycling resources, stabilizing human population growth, and eliminating poverty.

Developed nations would need to take a lead in this reversal process. Success in this endeavor, these scientists emphasized, requires a great reduction in violence and war, with the resources that are devoted to the preparation and conduct of war diverted to meet these new challenges. They also emphasized that it was only if all the world community came together — including scientists, governments, business, industrial, and spiritual leaders — that this could become possible.

It has been twenty-five years now since the report was issued. Many similar warnings have been issued since by scientists of every caliber, as well as by indigenous people speaking directly for Mother Earth. These warnings have been for the most part ignored. Environmental conditions have continued to deteriorate exponentially, a whole lot of new terms such as fracking, chemtrails, electrosmog and depleted uranium have been added to the mix, and we have continued to spiral down a dangerous path towards planetary destruction.

Nuclear accidents such as Chernobyl and Fukushima have contributed to this specter of death and destruction. Nuclear radiation from Fukushima has been drifting with the winds across the Pacific through the entire west coast of the Americas. Dolphins, whales, turtles and seals have been washing ashore by the thousands. Fish and other forms of sea life have been dying by hundreds of millions. An entire vast region of the Pacific Ocean has become a dead zone, where even the seafloor no longer supports life.

We are no longer dealing with separate environmental issues such as water pollution, topsoil depletion, climate extremes, desertification, deforestation, species extinction, toxic waste, nuclear radiation, atmospheric acidification, electromagnetic smog, and holes in the ozone. We have come to a point where the combined effects of long-standing planetary imbalances are creating unprecedented repercussions that we could not have imagined.

This brings us to the phenomenon known as global warming. Although there are natural cycles of warming and cooling that our planet has experienced for millions of years in its geological history, most people today associate this term with anthropogenic, or human-made, global warming, specifically referring to the deleterious effect of carbon emissions in our post-industrial civilization on everything from record high temperatures to catastrophic hurricanes to the death of coral reefs to the extinction of species. For many who truly care for the Earth, this theory has become a politically correct rallying call for saving the planet, confronting all the varied ills of a decadent civilization under a single banner of protest.

Where and how did this theory originate, and what does it signify for us in these troubled times?

In 1824, Joseph Fourier first came up with the idea of the greenhouse effect. The Earth gets energy from the Sun in the form of sunlight. The Earth's surface absorbs some of the energy and heats up. But much of this energy is also absorbed in the atmosphere, and directed down to the Earth. The more greenhouse gases there are in the atmosphere, including water vapor, nitrogen. oxygen, methane and carbon, the warmer the effects would be.

The hockey stick theory, first presented by Michael Mann and later popularized by James Hansen and Al Gore, focuses on one of these greenhouse gases, carbon dioxide, and the idea that the temperature of the Earth is rising exponentially in direct proportion to the rise of carbon-dioxide emissions, claiming that this has become the most important threat of these times to human and planetary survival.

Outlining the connection between global warming, carbon dioxide emissions, and irreversible climate changes such as extreme weather, floods, droughts, hurricanes, and epidemics, they assert that we need to seriously address these issues as a global community before it is too late. They propose drastic cutbacks in carbon emissions in order to combat these issues.[1]

There was a time when anyone who had an environmental conscience believed that greenhouse emissions were the enemy, and that we needed to do everything in our power to reduce these emissions. But what if the carbon emissions theory is at least partially misguided, and we need to be focusing our resources elsewhere?

A fierce debate has been going on over the past few years between proponents of global warming and between an ever-growing number of scientists who feel that global warming is a natural cycle and has very little to do with carbon dioxide emissions.

The entire premise of the hockey stick theory, that the Earth is significantly hotter since post-industrial times, is being questioned. In fact, says Professor Gernot Patzelt of Innsbruck University, the Earth has been at least this hot for over two thirds of the past ten thousand year period.[2]

Many of those who question this theory claim that much of the global warming research has been funded by governments and corporations for whom this represents a golden opportunity to reap unimaginable profits from the imposition of carbon taxes, and that there is no clear evidence that global warming and greenhouse emissions are related.[3]

A single volcanic eruption could put as much carbon dioxide into the atmosphere as an entire generation of gas guzzling vehicles, they point out. And even if greenhouse gases were responsible for a small percentage of global warming, the release of underground methane gases from practices such as fracking pose far more a threat than carbon dioxide emissions.

Many climate scientists do acknowledge that there is a warming trend going on, but rather than atmospheric warming from greenhouse emissions, they report that the Earth is simply responding to changes in the natural cycles of our Sun. Indeed, many of these scientists claim that a short-term cycle of global warming has ended, and we are now entering a period of global cooling instead, a theme we will further examine in the next few chapters.

Major world summits have taken up this debate, with endless squabbling about who is responsible, who is wrong, and who is right. The truth is that both sides may be right. The climate extremes and global environmental crisis we currently face is both manmade and natural. Global warming and global cooling are happening simultaneously. Summers are hotter; winters are cooler. The oceans are warming in certain parts of the world, cooling off in others. Coral is dying in response to warmer ocean climates and changing pH levels. Plankton is dying due to oxygen starvation and nuclear radiation.

Perhaps it is time to look not at isolated factors but at climate change as a whole. The seemingly separate phenomena are all interlinked in the big picture, as I attempt to describe in the following chapters. Whatever the outcome of the global warming or global cooling debates, these are not the only factors reflecting a world out of balance. And if it turns out that greenhouse gases are not as significant a factor as we had once thought, it still does not justify the continuing onslaught of pollution, deforestation and chemical destruction of our beautiful planet.

It is undeniable that the Earth is dangerously deforested, irradiated and polluted, that more species are going extinct every day, and that extreme climate changes have begun. We have a responsibility to the Earth and to future generations.

This is not just about science, nor is it just a political issue; rather it is the greatest survival challenge our civilization has ever faced. Whether manmade or natural, we find ourselves in the midst of a profound environmental and spiritual crisis. We need to find a broad enough platform of truth on which all camps can come together.

In the past hundred years we've already wiped out a tenth of all species on Earth, and species are becoming extinct at the rate of one every hour. A full third of all known species could disappear by the middle of this century.

This is an extinction level comparable to what happened 67 million years ago when a huge meteor struck the Earth, causing a chain reaction of events that obliterated half of our planet, including the dinosaurs. Or, closer to our own time, when huge numbers of large mammals, including wooly mammoths, giant sloths, dire wolves and saber-toothed tigers were wiped out in a cataclysmic event some 12,000 years ago. This next extinction could well include our own fragile species. As Chief Seattle put it, *"Contaminate your bed and you will one night suffocate in your own waste."*

From the global warmist perspective, a report, Meeting the Ultimate Challenge was put together in 2005 by a task force of senior politicians, business leaders, and academics around the world, intended as a guide for policy-makers in every country. In ten years or fewer, it indicated, global warming will have reached a point of no return.

The report even puts a figure on it, citing that a two degree Celsius increase over the average world temperatures prevailing in 1750 would irretrievably tip us over the edge, inducing widespread crop failures, water shortages, pestilence, death of forest and coral worlds, melting ice sheets, and switching off the Gulf Stream.

Those ten years are up. We have gone beyond the relatively benign and reversible problems we once faced. Whether we blame it on global warming, global cooling, human interference or natural cycles, we have already begun to see long-reaching, irreversible, and catastrophic Earth changes begin. Is another extinction level event on its way?

Dr. Guy McPherson, professor emeritus of Natural Resources, Ecology and Evolutionary Biology at the University of Arizona, does not have any qualms about this.[4]

Referring to the human species, he very sedately tells us that we have already crossed a planetary tipping point, and that there is nothing we can do to reverse this anymore. Basing his research on the global warming paradigm, he refers to a series of inter-connected feedback loops which are rapidly converging now, and gives humanity five to ten years before we become extinct through intensified and abrupt climate changes.

A major factor in his consideration is a vast 5-gigaton methane bubble trapped beneath the Arctic ice that is close to being released as ice caps melt. This would suddenly and drastically raise global temperatures by several degrees, leading to a chain of dire consequences.

McPherson is not alone in this view. Climate scientists and academicians such as Paul Beckwith, Jason Box, James Hansen, Michael Mann, Henry Pollack, Camille Parmesan, David Orr, and Peter Wadhams concur, some of them even claiming that McPherson is much too conservative in his estimates.

The good news is that Earth herself will survive, however, says McPherson, and continue to evolve, even if she has to start all over with microbes and bacteria.

CHAPTER 14
THE GREAT CONVEYOR BELT

Given this dire prognosis, what does global warming mean from the perspective of Gaia as a self-regulating super-organism? For most of human history, our species has evolved within a relatively narrow margin of tolerable temperatures. This has allowed for the proliferation and survival of our species. What would happen if the temperature of our planet suddenly increased or decreased beyond these limits? Indeed, since all species on Earth are woven together in an interdependent web of life, what would happen to the rest of planetary life?

I referred earlier to the current debate between those who believe that global warming is a man-made phenomenon and that we need to drastically reduce carbon dioxide emissions to curb this versus those who minimize the threat of global warming, claim this has very little to do with human interference, and emphasize that we need to refocus our attention and prepare for global cooling instead.

For instance, many are becoming aware of the geo-engineering phenomenon commonly known as chemtrails, where barium, aluminum and strontium compounds are being sprayed across the skies without popular consent, ostensibly in an attempt to reverse global warming. Not only is this a serious health hazard for human, plant and animal populations across the globe, but is accomplishing the exact opposite of what is needed at this time if indeed a global cooling cycle has begun.

Many climate scientists are questioning the relationship between carbon emissions and global warming, while others assert that this trend has now ended. While McPherson's prognosis, based on the global warming hypothesis, may appear grim, what if there were another trend currently counteracting this one? While summers continue to register record high temperatures, winters are getting increasingly colder, which leads some to argue that we are in for a period of global cooling instead of global warming.

Whether a consequence of human-made greenhouse gases, or an outcome of natural evolutionary cycles, or both, the fact remains that the temperature on the surface of the Earth has warmed approximately 0.5 degree Celsius on the average within the past few decades and until around 2007. At first sight this may not seem like much. But considering the intricately linked ecosystem of Gaia, this is a significant increase.

Imagine your own body's thermostat no longer set at its normal temperature of 37 degrees Celsius. Anyone who has had a fever knows what it's like to feel weakened, foggy, and sick, alternating between irritation and lethargy, cold chills and hot flashes.

The Earth has a variety of negative feedback loops to keep her temperature stable at the optimum rate for the proliferation of life. Consider what would happen if this thermostat were to be forcibly jammed over a long period of time.

Some global warming theorists claim that this rise in average global temperatures is enough to set a whole chain of events into motion, eventually even ending up creating a cycle of global cooling. How does this work? Consider one of the feedback loops of Gaia: Tropical oceans, being warmer, release more water to evaporate into the air, leaving them saltier. This moisture moves towards more temperate and polar climates where it precipitates as rain. This fresh water joins the sea, making it more diluted in these regions.

What oceanographers have discovered in the last twenty years is a vast underwater ocean current that is responsible for keeping temperate climates warm and tropical climates cool. Referred to as the Great Conveyer Belt, this vast undersea current is forty times the size of all Earth's rivers combined. The motor that keeps this belt running is the salt gradient between different regions of the world's oceans.

Warm water rises and cold water sinks. Likewise, water containing greater concentrations of sea salt is relatively heavier than water with lower concentrations of sea salt.

Warm water from the equatorial regions moves along the surface of the ocean towards the northern Atlantic regions. As it moves it shares this warmth with the landmasses it encounters, notably the western part of northern Europe and the eastern part of the North American continent.

As the warm ocean current moves north it gradually cools off. As it cools off, it begins to sink. At some point south of Greenland, cold Arctic currents from the north join what is known as the Gulf Stream from the south, pushing this massive stream of water deep below the surface to begin its return journey south.

This Great Conveyor Belt is one of the feedback loops of Gaia, and has, for thousands of years, maintained the relatively stable and moderate temperatures in Northern Europe and the east coast of America. Siberia, which is on the same latitude as England, is much colder because it does not experience the benefit of this warming action of the Gulf Stream.

Thom Hartmann reports in his book The Last Hours of Ancient Sunlight: Waking Up to Personal and Global Transformation, that the Great Conveyor Belt has been sputtering lately. It has already stopped and started a few times within the past ten decades. Once it shuts down completely, it is likely to remain shut down for anywhere between a few hundred years and a few thousand years, initiating a new Ice Age.

What is happening now to change this? As mentioned earlier, this great underwater conveyor belt is powered by the salt gradient. It depends on the higher concentrations of salt in the northern seas to drive the ocean currents down and southward.

Whether due to greenhouse emissions, as global warmists claim, or because of increased volcanism in the Arctic regions, as global coolists assert, a rapid meltdown of polar ice caps began taking place a few decades ago. Large pieces of Arctic ice caps broke off and started floating south. As these freshwater icecaps melted, it decreased the salinity of the water in the polar regions, in turn slowing down the pump feeding the Great Conveyor Belt.

Increased precipitation of rain and snow, brought in from the evaporation of oceans in equatorial regions, further contributed to this drop in salinity.

It is possible that the massive amounts of undersea oil that leaked from the Gulf of Mexico spill, also significantly contributed to the slowing down of this Gulf Stream and the subsequent winter freezes.

Meanwhile, the reckless surge of fracking in various parts of the world has contributed to the release of vast amounts of methane trapped beneath the Earth's surface, creating not only an increase in wildfire activity which has burned thousands of acres of homes and forests across the planet, but also contributing to the greenhouse effect.

There was a discovery made on May 23, 2016, reported on the Dutchsinse YouTube channel, about a supervolcano in the Arctic Ocean even bigger than the Mount Toba supervolcano in Indonesia or the Yellowstone caldera in Wyoming. Could this be one of the major causes of Arctic ice melting?

Either way, it is not just the changes taking place in undersea ocean currents. Changes in atmospheric wind currents caused by warming in certain parts of the world, significantly contribute to cooling or flooding in other parts.

Investigative journalist George Monbiot, for instance, provides strong evidence to show how abnormal warming in certain parts of the world is directly linked with changes in atmospheric pressure gradients. Thus, the freezing winters experienced in much of Europe and North America over the past few years seem to be a result of icy winds from Iceland replacing the milder winds from the Azores, a trend that is likely to continue in years to come.[1]

All things are connected. We cannot rightly understand either global warming or global cooling in isolation, but as symptoms of the same underlying planetary phenomena. The cycles are alternating faster and faster. What happens next?

CHAPTER 15
ICE AGES

An understanding of Gaia's feedback loops, such as the Great Conveyor Belt and the movement of atmospheric pressure gradients, provides clues for how warmer temperatures in some parts of the world might lead to cooler temperatures elsewhere. But what if there were larger forces at work than what we currently understand? What if the driving force behind Ice Ages is not related so much to terrestrial factors such as the Great Conveyor Belt, but rather to solar and galactic cycles?

In the 1970s scientists from CLIMAP (Climate Long-Range Investigation, Mapping, and Prediction) were examining deep-sea cores going back 500,000 years. They were startled to find that in the last half million years, ice ages had begun and ended abruptly. They were also amazed to discover that ice ages have come and gone on a regular basis, much more frequently than anyone previously had realized, and directly correlated to fluctuations in Earth's orbital patterns around the Sun known as the Milankovitch cycles.

They reported that we have had major Ice Ages coming and going every 23,000 years, and minor Ice Ages every 11,500 years. They also tell us that the last Ice Age ended between 11,400 and 11,500 years ago. Since then we have been in what's called an interglacial period, a gradual melting of ice and a movement towards warmer temperatures. However, this trend seems to be ending, and we are due to enter a major Ice Age any day now.[1]

In the last chapter we examined McPherson's assertion that humanity has perhaps ten years left before we go extinct. I don't wish to minimize our current crisis, but I believe that this is only part of the picture. McPherson and many of these climate scientists are basing their prognosis on the effects of global warming without taking into account an even more profound cycle of global cooling.

Climate scientists such as John Casey, David Hathaway, Dan Britt, Valentina Zharkova, Habibullo Abdussamatov, Lawrence Pierce and Zbigniew Jaworowski have been exploring data pointing to a very different outcome. Unlike McPherson, they tell us that the warming cycle is over, that the Earth's climate is now cooling, and that we have crossed an equally irreversible, but opposite tipping point. They give us this same five or ten year window for an abrupt shift towards a mini ice age.[2]

Unlike global warming theorists who base their findings on temperature measurements taken at various times and places on Earth, many of these scientists correlate climate conditions on Earth with cycles of sunspot activity within our Sun.

Former NASA scientist John L. Casey, states that a short-term cycle of global warming spanning the past few decades ended in 2007, and that in accordance with a 206 year cycle, we have currently entered a 30 year period of solar hibernation.

The recent global warming period had nothing to do with man-made greenhouse gases, he emphasizes Rather, all major climate changes on the Earth are driven by solar cycles. In his books, *Cold Sun* and *Dark Winter*, he demonstrates how we are rapidly moving into a period of extremely low solar activity, which would initiate a cooling cycle on Earth, as previously experienced during a period known as the Dalton Minimum between 1790 and 1830 AD.

If sunspot activity were low enough, he warns, we could even experience a Maunder minimum, a much colder spell that initiated the dark ages in Europe, lasting from 1645 to 1715 AD. The average global temperature during this period could be 5 degrees colder than our current average, leading to a severely curtailed growing season, food shortages, social collapse, and a flood of climate refugees going southwards.

In fact, a mini ice age already began in 2015, he claims, which will last for at least three decades and possibly more, depending on which way sunspot cycles go in the near future. In his book, *Upheaval*, he tells us that the worst earthquakes and volcanic eruptions tend to happen when we are at the very bottom of a solar hibernation period, also known as a grand solar minimum.

Casey expects the grand solar minimum to bottom between the years 2021 and 2046 AD. This not only means extremely low temperatures worldwide, but also a spike of earthquake and volcanic activity during this time. For example, the 1815 AD volcanic eruption of Mt Tambora in Indonesia, took place at the bottom of the Dalton minimum, and was one of the worst in recorded history, blanketing the entire planet with ash. Interestingly, the solar wind is at an all time low right now, with almost no sunspot activity evident. The extent of this solar minimum will determine the intensity of volcanic and earthquake activity during this time.[3]

Dr. Valentina Zharkova teaches astrophysics and mathematics at Northumbria University in Britain. She confirms Casey's findings, and refers to an approximately 400-year cycle of solar cooling, which is perhaps a combination of two of Casey's 206 year cycles. She clarifies that the Little Ice Age which began in the 1600's included both the Maunder Minimum and the Dalton Minimum, and expects that global temperatures during the upcoming solar minimum cycle, lasting from around 2020 till at least 2053 AD, will begin to plummet severely.

Habibullo Abdussamatov of the Russian Academy of Science goes further, saying that we can expect deep cooling starting with the end of Solar Cycle 24 in the year 2014, achieving its greatest severity around 2055 AD. He fears that we will likely experience weather conditions close to the bottom of the Maunder Minimum, the most severe cooling period within the Little Ice Age cycle.

I would highly recommend David Dubyne's YouTube channel, Adapt 2030, for his excellent interviews and periodic updates on the incoming cooling wave and its effects on global climate. It also provides a historical perspective based on earlier cycles of cooling, which could be crucial in predicting and preparing us for the changes ahead. Grand Solar Minimum is another informative channel I would recommend.

Mini ice ages are governed by changes in the output of 11-year sunspot cycles. They have to do with peaks and ebbs in the intensity of solar radiation. Classical Ice Ages are based on much longer periodicities, and have to do with the so-called Milankovitch cycles, based on Earth's orbit around the Sun, and including factors such as orbital eccentricity, axial tilt, and the precessional cycle.

Milutin Milankovitch was a Serbian mathematician in the early twentieth century. After a detailed analysis of interlocking planetary cycles, he came up with an explanation for the forces which drive ice ages. Orbital eccentricity is a 105,000 year cycle. The Earth's orbit around the Sun is not perfectly spherical but slightly eccentric, determined in part by the gravitational attraction of larger planets such as Jupiter and Saturn. The further away from the Sun we are the colder we get.

Axial tilt is the second of these three cycles. The tilt of the Earth is currently fixed at about 23.5 degrees, but this is not always so. Over a period of 41,000 years, our planetary axis fluctuates between 22.5 and 24.5 degrees. A lesser tilt leads to warmer winters and cooler summers, resulting in more precipitation in the winters along with shorter cooler summers, causing more snow to accumulate, leading to the long-term buildup of ice.

The third factor is precession, which refers to Earth's slow wobble as it spins on its own axis, causing the North Pole to slowly change its directional axis across the entire zodiac over the course of 23,000 years.

The precession of the equinoxes theoretically takes 25,800 years to trace a circle through the zodiacal signs. However, because the Earth is itself orbiting the Sun during this time, the combined effect produces a cycle of 23,000 years.

This precessional cycle, regarded as the pacemaker of the ages, is the primary indicator for Ice Ages. It breaks down into two cycles of 11,500 years each, one marking its way above the galactic equator in Earth's precessional journey, and one below, with the crossing points indicating times of galactic entrainment capable of initiating intense changes on our planet.

In addition to the precessional cycle, eccentricity and axial tilt also contribute significantly to the severity and duration of Ice Ages, the combined effect of which leads to shorter, 11,500-year periods of cooling and warming within longer and more intense Ice Age cycles. As we will see later, these crossing points are also linked with periodic reversals of Earth's geomagnetic field.

Based on the Milankovitch cycles as well as his own independent research, renegade scholar and science explorer, Randall Carlson, tells us that the relatively warm interglacial period that we have experienced in the past 12,000 years may soon be ending. Our current interglacial period is a warm interlude within a longer ice age period, he says, and once this interlude ends, we could once again experience not just a mini ice age, but epic Ice Age conditions, with average global temperatures 5-10 degrees below what we are experiencing today, and with glaciers covering much of the northern land masses.

The switch between a warm age and an ice age usually happens instantaneously, taking anywhere from a few months to a couple years. It is usually preceded by a few years of increasingly unpredictable and violent weather. It is also preceded by a period of global warming caused by underwater volcanism. Is this what we have been experiencing in the past few decades, incorrectly diagnosed as the result of greenhouse emissions?

Could it be that we have already entered the first stages of a mini ice age, and that a bigger classical ice age might be following close upon its heels?

Robert Felix is a science researcher and author of *Not by Fire but by Ice*. Regarding the advent of a classical Ice Age, he warns, *"The next ice age could begin any day. Next week, next month, next year... it's not a question of if, only when. One day you'll wake up ... buried beneath nine stories of snow. It's all part of a dependable, predictable cycle, a natural cycle that returns like clockwork every 11,500 years."*[4]

The conditions for the next ice age already exist on the planet, says Felix. The magnetic poles are fluctuating wildly and could reverse at any moment. It is already cold enough in northern regions to initiate ice age conditions. All it would take is increased precipitation — a requirement currently met by the cataclysmic flooding and record snows in different parts of the world.

Our planet is experiencing a cooling phase rather than a warming phase, Felix asserts. The basis for most claims about global warming is the melting of glaciers in the Arctic. But this is because of underwater volcanism in that area. Contrary to what is reported in the mainstream media, he says that glaciers and sea ice shelves are actually growing in many parts of the world, including Antarctica!

The increase of underwater volcanism around the world is the direct precursor for an ice age, continues Felix. It heats up the ocean causing more evaporation, which in turn creates more moisture and precipitation in the air. Combined with cooler temperatures from volcanic ash in the atmosphere, this process initiates an ice age.

In *The Survival of Civilization*, authors John D. Hamaker and Don Weaver demonstrate how carbon dioxide build-up leads not to global warming, but to rapid cooling of the globe, triggering accelerated change — from interglacial periods with relatively hospitable warm climates, to major ice age conditions — with most of humanity's crop-growing regions rendered extremely cold. Based on studies of pollen deposits, they conjecture that this transition from warm to cold conditions could happen in a matter of one or two decades. When they wrote their book thirty years ago, they believed that the new ice age could be upon us by 2000.

Unfortunately, the closer we get to any tipping point, the faster we move into a descending spiral. As icecaps melt, less sunlight is reflected back, which in turn allows more heat to be retained. Similarly, once an ice age begins, temperatures drop rapidly, and more landmass is permanently covered with ice and snow, which reflects back the Sun's light, lowering temperatures even further.

As much as we might like to blame our politicians and industrialists, it seems to me that greenhouse gases are a rather insignificant source for the climate changes we currently face. Humans certainly contribute to a world out of balance through endless pollution, deforestation, fracking, nuclear radiation, destruction of species, and reckless use of planetary resources, bringing us closer and closer to a dangerous tipping point. But as Casey, Carlson, Felix and others have demonstrated, there are other factors responsible for climate changes as well, as evidenced by the regularity of previous Ice Ages.

One of these factors is volcanism.

CHAPTER 16
VOLCANISM

Volcanic activity is another feedback loop for keeping Gaia within a precise temperature range. As the Earth heats up, it triggers sensitive pressure zones in the magma and lower crust. This sets off volcanic eruptions. Ash from volcanic eruptions blasts into the atmosphere, darkening the Sky, cutting off light from the Sun, and thereby cooling the Earth.

In the time of its first creation, Earth was a ball of superheated gases. As it cooled, the gas condensed to form fiery fields of lava within a layer known as the mantle. As the lava further condensed, it formed the crust of the Earth, the outer skin of Gaia. Over the course of millions of years, the crust of the Earth continued to expand due to centrifugal forces inside the Earth, while comets rained down from the Sky bringing dirty ice, which ultimately formed into oceans of water.

Ultimately, recognizable landmasses and oceans were born, all the while changing according to the shifting of tectonic plates floating over fields of magma in the lower crust.

There is a very interesting YouTube documentary presented by Neal Adams titled *The Earth is Growing*. Through animated computer simulations, Adams supports evidence earlier presented by Australian geologists Samuel Warren Carey and James Maxlow to show very convincingly that the Earth was originally a much smaller sphere with a single unbroken land-mass.

As Earth expanded in diameter over millions of years, cracks and folds started appearing within this landmass, much like a pinecone cracking open when it matures. Water, entering the planet's atmosphere through a rain of icy comets, began to fill these cracks, and became the oceans. Folds became mountains.

This activity of expansion also left enormous cavities within the Earth, which, incidentally, tends to support what is known as the Hollow Earth Theory. This fascinating theory postulates that due to centrifugal forces most of the molten mass of the Earth was pushed towards the outer circumference as Earth cooled and expanded, leaving an enormous empty space outside the fiery hot central core of the Earth.[1]

Meanwhile, as we look at the surface of the Earth, it is interesting to compare the western coastline of Africa with the eastern coastline of South America, and notice how easily they could have once been connected. Adams demonstrates in his animated model that the same applied for the Pacific spread as well, except on a much larger scale. The continents did not separate by merely sliding apart from each other but rather by mutually expanding outwards from the inner core and mantle of the Earth! [2]

According to this theory, the oceans developed as the Earth expanded. The deepest ocean valleys are the youngest portions of the Earth's crust and are where the last cracks appeared on the expanding Earth.

Scientists tell us that there is a ring of fire underneath the ocean beds, circling the Pacific Ocean and including portions offshore from Japan, Indonesia, and Tasmania, then around near the west coast of South America and all the way up off the coasts of California and the Hawaiian Islands. A large chain of underwater volcanoes sits astride this ring of fire.

The ring of fire corresponds to the youngest regions of the Earth. It is related to this most recent burst of large-scale creative activity in the current incarnation of our planet, as Adams theorizes.

Volcanism

Scientists are now telling us that human interference with nature is a relatively insignificant cause of global warming. The Earth is not simply heating up from the surface down, as would be the case if pollutants in the atmosphere were the main culprit. Rather, this planet is heating up from the core outwards. We will examine deeper reasons for this in later chapters, but for now, let us focus on the relationship between the heating of Earth's core and the volcanic ring of fire.

Scientists speculate that there is a cyclical relationship between core activity within the Earth and volcanic activity. Volcanoes along the ring of fire draw their heat from the layers of magma beneath the crust, and are in turn directly affected by activity inside the core of the Earth. Core activity in the Earth is, in turn, connected with cycles of change within the Sun, and with the Milankovitch cycles representing Earth's orbital patterns.

Volcanoes are classified as extinct or dormant depending on how long ago they last erupted. All volcanoes are potential hotspots, however, and if the Earth's core warmed up enough we would see a significant increase of explosive potential in both classifications. Residents of Washington state well remember Mount St. Helens and the massive cloud of volcanic ash that was released into the atmosphere when she erupted. What if the effects of such an event were to be multiplied many times over in years to come?

According to seismologist Michael Mandeville, along with climatologists at the Russian National Academy of Sciences, overall volcanic activity has increased 500% within the past hundred years, and is increasing exponentially.

Just as earthquakes are classified on a logarithmic Richter scale from 1 to 12, volcanoes are classified on a logarithmic scale measuring 1 to 8 — a Volcanic Explosivity Index (VEI). VEI 8 volcanoes are referred to as super-volcanoes, and are thousands of times more powerful than regular volcanoes such as Mount St. Helens, which was a VEI 5.

Super-volcanoes are capable of plunging the Earth into a volcanic winter, causing massive disruption of human life and possibly large-scale extinction of species. One such super-volcano erupted on the Mediterranean island of Santorini, close to present-day Crete, around 1500 BC. It was known as Thera in those days.

In that VEI 6 eruption, thirty cubic kilometers of volcanic ash spewed across Turkey, Greece, and Egypt, causing the sudden, sharp decline of the Minoan and Egyptian civilizations. Research by biblical scholars such as Graham Phillips (*Act of God*) indicates that this was the historical event precipitating the well-known biblical story of Moses and the ten plagues.

Many such super-volcanoes exist, most of them underwater either bordering a tectonic plate or following the ring of fire. The best known among them, however, is situated directly in the middle of a continent.

Most Americans are familiar with Yellowstone National Park. What most people are less familiar with is that Yellowstone is one of the biggest super-volcanoes existing on Earth, with a caldera 50 kilometers long and 20 kilometers wide.

As Lawrence E. Joseph reports in his book Apocalypse 2012: An Investigation Into Civilization's End, the periodicity of the Yellowstone caldera seems to be between 600,000 and 700,000 years. The last major eruption was 640,000 years ago. He cites recent geological surveys revealing that this entire caldera has risen about three-quarter meters since 1922, is filling up with magma, and getting ready to explode again.

Soil temperatures in some parts of Yellowstone National Park reached 200 degrees Celsius in 2003. Sections of the park have been inexplicably closed off to the public. A nearby river starting boiling in 2015. Although there is no way of knowing when the next eruption could take place, we do know that a growing number of scientists have been suggesting that the likelihood of volcanic eruptions is increasing in response to core activity within the Earth and Sun, and that this activity grows especially intense during times of magnetic reversal, a phenomenon which we will explore in detail later.

If Yellowstone explodes, it would be in the range of a VEI 8. *"It would be extremely devastating, on a scale we've probably never even thought about,"* says Robert Smith, a geologist at the University of Utah. Estimates of its explosive force range up to the equivalent of one thousand Hiroshima type atom bombs per second! It would be capable of plunging the entire western hemisphere of our planet into a volcanic winter from which our current civilization would never recover.

Unlike other super-volcanoes, which reach down as much as 1,800 miles beneath the surface of the Earth to the boundary between the core and mantle, Yellowstone reaches a depth of only 125 miles. Rather than being fed by the heat at the molten core of the Earth, it feeds off the vast reserves of uranium and other radioactive material underneath the Earth in that area.

We are talking not just about a volcanic winter here, but the potential of a nuclear volcanic winter.

We are aware of thirty or so large super-volcanoes around the world. There are likely many more. Most are underwater and, if they erupt, are capable of generating massive tidal waves, as we experienced in the tsunami of 2004. Increased volcanism is another link in a causal chain connecting Ice Ages, solar cycles, and even galactic cycles, as we shall explore in the next few chapters.

CHAPTER 17
SOLAR CYCLES

L ife could not exist on Earth without the heat and light
provided by our Sun. Yet these conditions have not always
existed. What distinguishes Earth from many of the other
planets in our Solar System is our ability to retain the heat of the
Sun due to the gradual growth of an atmosphere.

Another essential ingredient for life on our planet is the
existence of an electromagnetic field, which provides us with an
orientation in space and time and acts as a protective barrier
around the boundaries of the Earth. Without this magnetic
shield we would be bombarded by meteorites and cosmic dust
orbiting through the Solar System, as well as from the
fluctuations of solar activity known to us as solar flares and
coronal mass ejections.

There are cycles that govern this solar activity. One such cycle
that has generated significant interest in recent times is the
eleven-year sunspot cycle. Based on a magnetic differential
between the equatorial and polar regions of the Sun, the peaks of
a sunspot cycle are related to corresponding increases in solar
activity.

During the peak of a sunspot cycle, violent explosions on the
Sun, shoot photons and high-energy particles towards the Earth,
jolting our ionosphere and geomagnetic fields and potentially

downing power grids and satellites, thereby disrupting communication systems.

The intensity of a sunspot cycle is measured by the maximum number of visible sunspots — dark blotches on the Sun that signify areas of increased magnetic activity. The greater the number of sunspots, the greater the likelihood of major solar flares, or the even more violent coronal mass ejections in which the entire gas shell surrounding the Sun, known as the corona, gets ejected out into space.

Beyond the eleven-year sunspot cycle, however, there seem to be additional factors involved in the generation of extreme solar activity.

Some years ago, Sami Solanki of the Max Planck Institute for Solar System Research in Germany reported that the Sun was more violently active currently than at any time in the past 11,000 years. Since 1940, he claims, the Sun has produced an extraordinary number of sunspots, as well as solar flares and coronal mass ejections, which cannot be explained by the normal fluctuations of the eleven-year sunspot cycle.

This has changed dramatically in the past couple years, as we discussed in the previous chapter, with the Sun having entered a quiescent phase in 2015, which is likely to last the next 30 years or more.

Meanwhile, the Earth's magnetic field, which normally protects us from solar and cosmic radiation, has been inexplicably dwindling in recent years. Could all these phenomena be related? How do we explain what is happening? Astrophysicist Alexey Dmitriev of the Russian National Academy of Sciences in Siberia provides us with some very interesting clues.

Just as the Earth revolves around the Sun, the Sun travels its own orbit through our Milky Way galaxy. And just as the Earth has a protective electromagnetic shield surrounding it in its journey through the Solar System, so does the Sun have a protective electromagnetic envelope surrounding the entire Solar System in its journey through the Milky Way.

The magnetic field surrounding our Solar System is known as the heliosphere. The outer edge of this heliosphere is known as the heliopause, and is composed of highly charged layers of plasma, defined as minute electrical particles. As the Sun travels through the Milky Way, this plasma acts as a shield, repelling and pushing away galactic cosmic rays and particles of interstellar dust.

What Dmitriev and his team have been observing is that the shockwave of plasma at the leading edge of this heliosphere has expanded ten-fold in size and intensity during the past two or three decades. It has grown from about 3 AU (astronomical units) to 40 or more!

In his paper, *Planetophysical State of the Earth and Life,* Dmitriev states, *"This shock wave thickening has caused the formation of a collusive plasma in a parietal layer, which has led to a plasma overdraft around the Solar System, and then to its breakthrough into interplanetary domains."*[1]

Many in my generation were brought up on Star Trek. Imagine the Starship Enterprise entering a region of space saturated with meteorites and cosmic dust. Imagine Captain Picard commanding his crew to direct all power towards the shields in order to strengthen the force fields around the ship and to repel the incoming stardust.

This is what is happening with our Solar System. Dmitriev asserts that deep space is not uniform. Composed of scattered space debris, some regions of space are more turbulent than others. We are entering into one such region of interstellar turbulence. The shockwave in front of the heliosphere has become larger and thicker as we entered this denser region of space where there are more particles to push out of the way.

Imagine now that the Starship Enterprise is facing an intense bombardment of cosmic stardust, and the overloaded shields are flaming hot. This is precisely what is taking place in our own Solar System. The interstellar turbulence of cosmic dust is overloading and breaking past the plasma shield, causing what Dmitriev describes as *"a kind of matter and energy donation made by interplanetary space to our Solar System."*

In other words, things are heating up. By analyzing data from the Voyager satellites out in deep space, Dmitriev and his team confirmed that this rise in temperature was affecting not only our Sun but also every other planet in the Solar System. The interstellar turbulence is also causing a whole range of other measurable effects in our Solar System — the growth of planetary atmospheres, the increasing brightness of various planets, changes in magnetic fields, and shifting rotational poles.

Here on Earth we are experiencing the impact of this interstellar dust, both directly as a result of this energy donation, and indirectly through changes within the Sun. The anomalous activity of the Sun and its unpredictable effects on the Earth could well be related to this interstellar region of turbulence we are currently entering.

Dmitriev expects that we will remain within this turbulent shockwave for at least the next 3,000 years. He believes also that many of the extreme and inevitable climate changes could begin very soon — not in centuries or even decades, but in a few short years.

CHAPTER 18
GALACTIC SUPERWAVES

We are ready now to explore what could be one of the primary mechanisms for changes happening on our planet today. This mechanism includes a likely explanation for Dmitriev's observations of increased cosmic dust activity surrounding our Solar System, as well as for cycle of magnetic reversals and corresponding Ice Ages on Earth. This is the Galactic Superwave Theory, proposed and developed by astrophysicist Paul LaViolette.

I was privileged to meet Dr. LaViolette at a new sciences symposium in Las Vegas back in 2001. My first book, *Doorway to Eternity: A Guide to Planetary Ascension*, had just been released, as had his own groundbreaking book, *Earth Under Fire: Humanity's Survival of the Ice Age*, and we found ourselves sitting next to each other at our respective booths. I found him gentle and unassuming, in sharp contrast to the disturbing implications of his theories.

Over the years, we have developed a quiet friendship and have had some fascinating discussions on his theories of continuous creation and galactic superwaves. I am grateful to Paul for helping me fine-tune my understanding of these themes, and for serving as a science advisor to me as I prepared this manuscript for publication.

Galactic core explosions are a recent discovery. During the early '60s astronomers noticed that the cores of spiral galaxies periodically become active. When active, they spew out enough cosmic ray and gamma ray energies to equal millions of supernova explosions, which are the explosions caused by individual stars in the last phase of their life cycle.

They assumed that these active states happened fairly infrequently, about once every ten million years or more. LaViolette proved otherwise. Using astronomical data and the study of ice core samples, he hypothesized that these galactic core explosions were much more frequent than anyone previously realized.

Beryllium-10 is an isotope of Beryllium gas produced by the effect of galactic cosmic rays. In the course of his research, LaViolette discovered a significant concentration of Beryllium-10 in ice core samples dating 10 to 15,000 years ago. This supported his earlier hypothesis based on astronomical data, as well as his experiments with Nickel and Iridium indicators in ice core samples, and he conjectured that this was caused by recurring galactic core bursts following a periodicity of approximately 10 to 15,000 years.

This periodicity was later confirmed by ice core data analyzed by Grant Raisbeck, which showed the presence of regular Beryllium-10 spikes ranging over the past 140,000 years, thus providing direct proof for periodic bursts of cosmic ray activity emanating from the galactic center.

LaViolette asserts that each galactic burst could last several hundred or even thousands of years. In our own galaxy, the last major event took place approximately 10 to 15,000 years ago. Additionally, astronomical observations indicate that there have been fourteen minor expulsions of ionized gas within the past 6,000 years.

LaViolette coined the term galactic superwave to refer to this cosmic ray bombardment. He suggested that cosmic ray particles emanate from the super-dense center of our Milky Way galaxy and rapidly move out in spherical waves to the outer limits of its spiral arms at nearly the speed of light. As this superwave moves out radially throughout the galaxy, unaffected by interstellar magnetic fields, it generates a wave of gamma rays in front of it.

Just as our hearts send out blood through arteries and veins in a steady rhythm, so our galaxy has its own heartbeat, sending out energy and information in the form of cosmic and gamma rays throughout its galactic body, with one major pulse approximately every 10 to 15,000 years, and a number of minor pulses in between.[1]

At this point I am going to take the liberty of assigning a value of approximately 12,000 years to this galactic heartbeat. Although LaViolette does not try and pin this down quite so specifically, I believe this could prove to be a starting point for further explorations and I am curious to see where it may lead.

LaViolette's galactic superwaves and Dmitriev's dust cloud model seem to have a complex relationship with each other. Dmitriev hypothesizes that as our Sun spins around the Galactic Center, we are currently entering a region of the galaxy highly saturated with interstellar gases, dust and debris. These interstellar dust clouds are usually held at bay outside the Solar System by plasma shields at the outer boundary of our Sun's magnetic field.

However, whenever a galactic superwave makes its periodic journey through our Solar System, trillions of high velocity protons emitted by the incoming cosmic and gamma rays cause these plasma shields to become overloaded, and eventually to fail, pushing these orbiting clouds of cosmic dust inside the heliosphere, initiating a cycle of solar and interplanetary turbulence.

Support for the dust cloud theory also emerges from the observations of astronomer Hartmut Aumann, who suggested in 1988 that the Solar System is surrounded by a dust envelope five hundred times denser than previously recognized. Observations by the Infrared Astronomical Satellite in 1984 further confirm that the Solar System is currently surrounded by cirrus dust clouds.

A couple decades ago, Markus Landgraf of the European Space Agency warned us that the heliosphere could be close to collapsing. He spoke of increasing turbulence and flaring within the Sun, leading to a disturbance of the heliosphere, causing an eventual collapse of the heliopause, allowing vast quantities of cosmic dust to pour in. The same scenario could take place if the Sun enters a cooling phase, as seems to be taking place now.

LaViolette emphasizes that the effects of the galactic superwave on the Sun and on the Earth's climate are not so much due to the cosmic and gamma rays themselves, but to the cosmic dust, cometary bodies, and ice chunks of all sizes that these galactic rays transport into the Solar System. This invasion of interstellar debris is capable of initiating a chain reaction of increasingly violent events on the Earth as well as on the Sun. This is the mechanism triggering the recurring ice ages, as well as the periodic weakening and reversal of Earth's magnetic fields, over the past three million years.

Indeed, this could also be the mechanism causing periodic meteorite strikes on Earth and other planets. The breakdown of the heliopause allows chunks of space debris from these galactic dust clouds as well as from the Kuiper belt and Oort clouds, to enter into our Solar System. Earth and the inner planets are normally protected by the gravitational giants, Jupiter and Saturn, with most of this space debris being drawn into their gaseous orbits.

However, it is possible for some of them to break past into Earth's orbit, resulting in the sort of meteor strike that we experienced 12,900 years ago, when according to Randall Carlson and Graham Hancock, huge crustal displacements and massive extinctions took place on Earth. A huge comet hit the polar ice during this time, instantly dissolving the polar ice caps, raising sea levels hundreds of feet, and carving massive canyons into the landscape. This spelled the end of the previous Ice Age, and the beginning of our current interglacial period.[2]

For the record, this is an area where researchers such as Robert Felix would disagree, stating that the catastrophic ending of our previous Ice Age resulted not from a cometary impact but from electrical turbulence and volcanic activity initiated by the recurring 11,500 year precessional cycle, the pacemaker of the ages, where strong electrical discharges from our alignment with the galactic equator stimulated extreme volcanic and seismic activity upon the Earth, resulting also in a geomagnetic reversal.

This would also tally with conclusions derived from the Electric Universe theory, as explored in later chapters.

Going back to LaViolette, what would be the likely effects of the next galactic superwave? As cosmic dust and debris bombard the Sun's surface, the magnitude of solar flares could increase, causing the Sun to engage in continual flaring activity and creating enormous coronal mass ejections. Simultaneously, as Earth's magnetic field weakens, even moderate flaring would have huge effects on the Earth's climate and communication systems. We will discuss this further in later chapters.

It is interesting to note, according to the ESA, that the amount of cosmic dust in our Solar System has increased three-fold since AD 2000. As a result of this increase, and corresponding increases in solar activity, communication networks, power grids, and satellite systems could be shut down permanently as immense magnetic storms rock the Earth. Floods, earthquakes, volcanic and tidal wave activity could follow. Immense heat could penetrate the Earth's atmosphere, burning forests, igniting volcanoes, evaporating oceans, and precipitating a new Ice Age.

If the cosmic dust incursion is sufficiently intense, says LaViolette, highly charged cosmic rays could be generated by the Sun, and the resulting solar winds could bombard the Earth's surface, generating immensely strong ring currents around the equator that could cancel and reverse the Earth's magnetic field in a matter of days. Meanwhile, as likely happened about 12,900 years ago, giant coronal mass ejections could once again blast through interplanetary space and envelop the Earth in a huge fireball, causing a mass species extinction.

Note that the extinction event explained by Randall Carlson as a comet strike and by Robert Felix as the result of precessional crossing, is interpreted by Paul LaViolette as an outcome of coronal mass ejections from the Sun. It could be that these are all related consequences of a major galactic superwave event between 12 and 13 thousand years ago, a category 4 on LaViolette's scale. Minor episodes, which occur more frequently, would not be as catastrophic, although still powerful enough to severely disrupt our current civilization.

What does it mean that the Sun is currently going through a quiescent phase, with sunspot activity diminishing to near zero levels? Could this be a lull before the storm, a prelude to the next blast of incoming galactic energies? John Casey does make a correlation between low sunspot activity and increased volcanic and seismic activity. I would be curious to explore the factors initiating the Little Ice Age in the 1600's, and whether this had anything to do with a minor galactic superwave bombarding our solar system at that time.

Of particular concern, cautions LaViolette, is the fact that cosmic rays travel close to the speed of light and so could strike our Solar System virtually without warning. Preceded only by the wave-flash from the initial explosion, we would perceive it as a bright bluish-white light having the appearance of a very bright star coming from the galactic core in the constellation of Sagittarius. We would also notice a bluish haze caused by the cosmic rays impacting the heliopause, but the center of it all would be this star-like phenomenon.

Interestingly, there is a Hopi prophecy about a blue star becoming visible in the days of purification preceding our journey into the Fifth World. Is this prophecy based on a memory held over from the last time this happened?

When asked about when he expects the next major galactic superwave volley here in our Solar System, LaViolette makes a conservative guess, stating a 90% chance it could arrive within the next four centuries. He concurs, however, with others that a catastrophic event might occur much sooner.

Although the galactic core seems to be in a quiescent state at the moment, LaViolette says this could change at any time. What would the passage of a galactic superwave look like? Are we in the first stages of this already? And how might this be related to changes in Earth's magnetic fields?

CHAPTER 19
GEOMAGNETIC
REVERSALS

There are two kinds of Pole shifts generally talked about — geographical pole shifts, and geomagnetic reversals. In this chapter, we will limit ourselves to a discussion of magnetic reversals.

The Earth is like a giant dipolar magnet with magnetic north and south poles that tend to wander from time to time, usually circumscribing the geographical south and north poles in a radius of a few hundred miles.

This magnetic field has existed for at least 3 billion years, although it tends to fluctuate wildly from time to time, and even reverse itself on a periodic basis.

The Earth's magnetic field is generated deep inside the planet. An inner core of highly pressurized solid iron is surrounded by an outer core of molten iron and nickel. They rotate at different rates, and the interaction between them creates a hydro-magnetic dynamo. Not unlike an electrical dynamo, this mechanism is what creates and sustains the geo-magnetic field.

In fact, seismologist Michael Mandeville explains that there is not a single magnetic field but a series of five interactive magnetic fields generated between core, magma and surface of the earth, and reaching out into the atmosphere and ionosphere, all combining together to produce an average field intensity. It is a complex mechanism that can be triggered by various influences including solar flares, galactic waves, astronomical influences, and other subtle factors.

Similar to the lines of force surrounding a bar magnet, the magnetic field of our planet not only runs through the Earth itself but also creates lines of force surrounding the Earth. These lines of force, which include the Van Allen Belts, act as a protective shield for repelling excessive solar and cosmic radiation, as well as meteorites and space debris.

The geomagnetic field is not static. A study of the geological record indicates many fluctuations as well as periodic field reversals. At present, the Earth's magnetic field seems to be weakening exponentially. It is the lowest it has been since the past 11,500 years, with a decline of 15-20% within just the past 200 years. Significant breaches in the magnetic field have been observed recently.

Geologists disagree about how long it takes between reversals or whether there is even a predictable period of time between reversals. They are also divided about how long it takes for a magnetic field to collapse and rebuild.

Those who believe in a uniform model of planetary evolution, implying that volcanic sedimentation in ice core samples is deposited evenly over the time between reversals, assert that the periodicity of these polar reversals could be anywhere from 100,000 to one million years. Andrew Jackson of Leeds University estimates that the last reversal occurred 750,000 years ago. LaViolette points to evidence for a more recent reversal called the Blake Event about 100,000 years ago.

On the other hand, geologists who tend to support a catastrophic model of planetary evolution, believing that sedimentation takes place erratically, based on a succession of extreme geological events, seem to feel that the poles reverse themselves much more frequently.

For example, as Robert Felix point out, the Gothenburg magnetic reversal occurred about 11,500 years ago, the Mono

Lake reversal 23,000 years ago, and the Lake Mungo reversal 34,500 years ago. In his book, *Not by Fire but by Ice*, he charts magnetic reversals and catastrophic events going back 127,000 years in perfect synchronicity with this 11,500 cycle of equinoctial precession. Ice ages usually commenced and ended during such reversals.

In *Reality Revealed: The Theory of Multidimensional Reality*, authors Douglas Vogt and Gary Sultan contend, using potassium-argon dating techniques, that magnetic reversals are spaced apart quite evenly in length. They assert that the periodicity between successive polar reversals is approximately 12,000 years, and that the most recent reversal was 12,000 years ago. They further state that this last polar reversal coincided with the ending of the previous Ice Age.

Like Robert Felix, they acknowledge the importance of the Milankovitch cycles in driving the recurring ice ages and magnetic reversals, but refer to a clock of 12,068 years based on studies of the Kabbalah.

Magnetic reversals follow a predictable natural cycle, and always begin rather suddenly. They are usually initiated in less than twenty years, and then continue on for a few hundred or a few thousand years, depending on various conditions. Once the magnetic polarity reverses it could either continue in its reversed polarity for the duration of the cycle, in which case it is known as a magnetic reversal, or else it could revert back after a few hundred years to its original polarity, in which case it is known as a magnetic excursion.

There are other researchers who have also linked this same cycle with historical events. For example, in their book *When the Earth Nearly Died: Compelling Evidence of a World Cataclysm 11,500 Years Ago*, D.S. Allan and J.B. Delair provide evidence of a global catastrophe 11,500 years ago, which they claim was related to Noah's flood. I suspect this may have been a retelling of the Atlantean story from a different cultural perspective.

Similarly, in *The Celestial Clock*, William Gaspar refers to a dominant ice volume collapse cycle, where the polar axis shifts every 11,500 and 23,000 years. The shift is accompanied by severe earthquakes and flooding.

Meanwhile, as mentioned earlier, Vogt refers to a precise 12,068-year clock that governs ice ages and magnetic reversals, and diligently details what a magnetic reversal could look like in our own time, the last phase of which would occur in the space of just a few hours. The reversal would begin with spikes in solar activity, which he predicts will occur sometime between September and December of 2046. This would be followed immediately by a magnetic reversal on the Earth, and then an ice age.

Many of these scientists and theorists share the same periodicity of approximately 11,500 or 12,000 years. Is there a relationship between galactic superwaves, magnetic reversals, and ice ages? Do they follow a sequence? If so, what is this sequence, and how does it work? If we experienced this before could it happen again? When might this happen and what could this mean?

Consider the following worst-case scenario. A minor or major galactic superwave passes through our Solar System. The intense bombardment of cosmic ray and gamma ray particles overload the plasma shields surrounding our Solar System, allowing thick clouds of cosmic dust to enter in. This initiates a cycle of increased solar activity. Intense flares and coronal mass ejections move from the surface of the Sun out through inter-planetary space.

At the peak of solar activity an immense cloud of plasma catapults out into space as solar lightning. As it reaches us, it overloads the Earth's geomagnetic field. The hydro-magnetic dynamo shuts down, during which time the protective magnetic field of the Earth briefly collapses, then reverses.

There is an additional possibility relative to magnetic reversals that I will mention here. There is a rule pertaining to electromagnetic dynamos known as the right hand rule. Imagine making a fist of your right hand with the thumb sticking up. If the direction of the thumb represents the North Pole, then the curl of the fingers represents the direction of spin. The stronger the magnetic field, the faster the spin. If the magnetic polarity reverses, the spin reverses. Some authors have applied this principle to the spin of the Earth and argued that a magnetic reversal would inevitably lead to a reversal in the Earth's direction of spin.

There seems to be some evidence for this possibility in ancient writings from various cultures testifying about cataclysmic times when the stars fell from the Sky, or when the Sun rose from the west rather than the east. Some have even pointed to such evidence as the depiction of the zodiac in ancient temples, such as Dendera in Egypt, where the sequence of constellations is drawn reversed.

Although LaViolette was one of the first to suggest a link between galactic superwaves and magnetic reversals, he feels that the possibility of the Earth reversing its rotational spin is extremely unlikely. I would welcome further dialogue on this issue.

CHAPTER 20
CRUSTAL DISPLACEMENT

E dgar Cayce, well known as the 'sleeping prophet', predicted a pole shift in 2001.

Fortunately for humans on this Earth, his prediction failed. Unlike magnetic reversals, a pole shift refers to a literal shift of the continents and oceans on the surface of the Earth, relative to the north-south polar axis.

I didn't think this was necessarily connected with magnetic reversals until I came across some research by Charles Hapgood, history professor at Keene State College in New Hampshire. In his book, *The Path of the Pole*, Hapgood plots the course of geographical pole shifts through the ages, asserting that the last pole shift happened between 11,500 and 12,000 years ago.

Here was that number again. I began to wonder if this could be yet another factor in a single periodic chain of events? Based on geomagnetic studies and Carbon-14 analysis, Hapgood arrived at the conclusion that 12,000 years ago, before the poles shifted to their current position, the North Pole was centered in the Hudson Bay region.

He also highlighted two other pole shifts in recent geological times, showing that the north pole was located in the Greenland Sea 55,000 years ago and in the Yukon area 80,000 years ago. Each period of shifting took 5,000 years, he claimed.

He referred to the mechanism responsible for this shift as Earth Crust Displacement or crustal plate displacement. The outer crust of the Earth floats upon a semi-liquid layer of magma. Under certain circumstances, the entire crust of the Earth can shift in one piece, like the loose skin of a tangerine orange.

This is a very different phenomenon from continental drift. Continental drift refers to continents moving away from each other, almost imperceptibly, in time. Crustal displacement refers to the whole crust of the planet, including all continents and ocean beds, moving together to a new position relative to the polar axis.

How does this happen? Hapgood wasn't quite sure, yet proposed that, as we enter an ice age, the increasing weight of ice in the polar regions would cause an imbalance in the spin of the Earth around its axis. Like a washing machine out of balance, this could create a shudder in the Earth's rotational orbit. Once a tipping point is reached, the entire outer crust of the Earth would shift to a new stable position.

In interpreting the geological evidence, Hapgood came to the conclusion that the movements to each new position were not cataclysmically fast but relatively slow, taking about 5,000 years to make the journey. He postulated that the new North Pole remained in place for about 20,000 to 30,000 years after each shift.

Another explanation for crustal plate displacements has to do with the conjecture that magnetic reversals would inevitably lead to a reversal of the Earth's spin, as discussed in the previous chapter. Such an event, were it to happen, would generate enormous tsunamis, hurricanes, earthquake and volcanic activity that could conceivably wipe out everything on Earth.

Although I do not personally believe that magnetic reversals necessarily lead to a reversal or shift of Earth's rotation, a crustal plate displacement would act as a cushion in softening the intensity of such an event, were it to happen. The same applies to cometary impacts and other unusual stresses on the body of the Earth.

Whether caused by unbalanced ice, magnetic reversals or cometary impacts, there is a safety mechanism involved in the process where the outer crust of the Earth disengages from the lower crust and mantle, sliding freely over a semi-liquid layer of magma, allowing it to shift positions in a relatively gentle manner even with all the tremendous forces being generated beneath it.

Author and spiritual teacher Drunvalo Melchizedek refers to the discovery that whenever the magnetic field of the Earth drops down to zero, then in about two weeks the upper mantle of the Earth would become liquefied, and the Earth would be free to spin in any direction it wants. At this point, he conjectures, the unbalanced weight of the Antarctic Ice would cause the entire planet to be crustally displaced to another location.[1]

The last time we experienced a crustal plate displacement, according to Hapgood, was 12,000 years ago. As I read this, my thoughts drifted to stories of Atlantis and its final cataclysmic sinking. If this was a real event, when and how did this vast landmass actually sink? Belgian author Patrick Geryl refers to The Egyptian Book of the Dead as translated by the Frenchman Albert Slosman. One of the oldest of Egyptian manuscripts, the book provides a date for this event at 9792 BC, roughly 11,800 years ago.

Authors Rand and Rose Flem-Ath were students and colleagues of the late Professor Hapgood. In their book *The Atlantis Blueprint*, they make a strong case that the remnants of Atlantis, once a thriving civilization similar to our own, are buried underneath the Antarctic Ice. Graham Hancock, in Fingerprints of the Gods, concurs. If this is so, it seems that the entire continent of Atlantis could have moved towards the South Pole as it sank during this last period of crustal displacement and related catastrophic activity.

For the record, this is an area where Hancock and LaViolette disagree. LaViolette argues that polar ice core results rule out the possibility of any latitudinal change of Antarctica.

Regardless of the final verdict, is the fate of Atlantis one probability for what lies before us in the very near future? Is this what the Maya and other indigenous people, prophets including Edgar Cayce, and scientists such as Paul LaViolette, Randall Carlson and Charles Hapgood, have been warning us about?

CHAPTER 21
COSMIC AGES AND TWIN SUNS

In my early research into ancient calendar systems the precessional cycle came up time and time again. Also known as the precession of the equinoxes, the cycle marks the apparent pathway of the north polar axis through the twelve constellations of the zodiac based on measurements taken from Earth at sunrise on the autumnal equinox.

Just as the Earth experiences seasons in its journey around the Sun, so the precessional cycle refers to zodiacal seasons. Both these seasons have their basis in the tilting of the Earth's axis. The tilt of the Earth fluctuates slightly from time to time creating corresponding changes in the length of the precessional cycle.

We cross the plane of the galactic equator twice during this precessional journey through the zodiac. Please remember we are not talking about a physical crossing of our Solar System across the galactic plane. We are simply referencing an alignment of the north polar axis with the central plane of the galaxy due to the tilt of Earth's axis.

This happens at the halfway points of each precessional cycle, and during these times we are more responsive to galactic energies. In this current cycle our passage through the galactic equator began in 1980 and ended in 2016.

Moira Timms, author of *Beyond Prophecies and Predictions: Everyone's Guide to the Coming Changes*, once suggested to me that energies streaming in through the galactic equator during these halfway points could be an alternate explanation for LaViolette's theory of galactic superwaves. When I confronted LaViolette with this explanation, he disagreed, stating that the galactic superwave is a physical blast of cosmic ray energy, not to be confused with the precessional cycle.

It does seem, however, that these could be an indirect relationship between these two factors. The timing of the cataclysmic events and evolutionary changes we have discussed in preceding chapters is initiated by a galactic heartbeat of approximately 12,000 years, which generates bursts of cosmic and gamma rays moving outward concentrically as galactic superwaves. We could be experiencing an inflow of these cosmic rays shortly.

On a secondary level, as indicated by Robert Felix, we also receive galactic energies from the galactic equator during the crossover points of the precessional cycle. Through a system of electromagnetic resonance, which we will examine in more detail later, the Earth's polar axis becomes entrained with the informational waves of a galactic heartbeat extending through the galactic equatorial plane. Could it be that precessional crossings serve as an early warning mechanism for incoming superwaves? And if so, does the fact that we have recently crossed the galactic equator mean that a galactic superwave could shortly follow?

One calendar system that apparently correlates with this recurring cycle is Sri Yukteswar's conception of the yugas, or cosmic ages. A highly accomplished yogi and seer, he was the spiritual teacher of Yogananda Paramahamsa and the direct disciple of both Lahiri Mahasaya and the immortal yogi Babaji.

In contrast to the much longer cycle of yugas in classical Hindu texts, which he considered flawed, Sri Yukteswar discussed the merits of a 24,000-year cycle of the yugas, during which time we pass through a sequence of events gradually leading us from an age of light into an age of darkness, and then back again.

There is a 12,000-year descending arc, followed by a 12,000-year ascending arc. Just like the Milankovitch half-cycle, this half-cycle of the yugas appears to closely match some of the cycles that drive evolution, including galactic superwaves, the ending or beginning of ice ages, and magnetic reversals on Earth.

I have experienced a strong inner connection with Sri Yukteswar and his immortal teacher, Babaji. As with the Mayan calendar, I feel that the cycle of the yugas is based in a galactic understanding of consciousness and time.

In his book, *The Holy Science*, Sri Yukteswar not only tells us the timing of these cycles but also hints at the astronomical reasons behind this:

"We learn from Oriental astrology that moons revolve around their planets and planets around their Sun; and the Sun, with its planets and their moons, takes some Star for its dual and revolves around it in about 24,000 years of our time – a celestial movement which causes the backward movement of the equinoctial points around the zodiac. The Sun also has another movement by which it revolves around a grand center which is called Vishnunabhi, which is the seat of the creative power, Brahma."[1]

The grand center that Sri Yukteswar refers to is the Galactic Center, source of the galactic superwaves that periodically ripple out through our own Solar System, which the Mayans referred to as Hunab Ku. But what is this dual star that he refers to, which apparently causes the precession of the equinoxes?

Astronomers have recently noted that our Sun's journey around the galactic center is not a simple orbit but rather a complex wobble that seems to indicate that our Sun might have a twin star that it circles around while they both simultaneously orbit around the galactic center.

Walter Cruttinden, in his book, *The Lost Star of Myth and Time*, conjectures that this twin star could actually be Sirius, and that the backward movement of our zodiac that scientists have previously explained as a result of Earth's axial tilt could equally be explained by the wobble created by these twin stars as they orbit around each other in their journey through the galaxy. This wobble makes a cycle of 24,000 years, just as Sri Yukteswar asserted, which Plato referred to as the Great Year, and what Milutin Malkovitch refers to as the evolutionary pacemaker.[2]

The idea of Sirius being a sister sun to our own Sun might be new for many people, but Cruttinden makes a convincing case for this based on astronomical calculations. Most stars are binary or tertiary systems, and so it is not so difficult to imagine them spinning around each other over vast distances as they jointly make an even larger orbit around the galactic center.[3]

These twin Suns wander through space in a highly elliptical orbit relative to each other, says Cruttinden, speeding up as they come close and slowing down as they drift apart. In terms of the cycle of yugas, our relative proximity to Sirius at any given time determines which yuga we happen to be in.

Sri Yukteswar indicates that we have passed out of the kali yuga, the age of darkness furthest from the light of our twin star, and are currently in a dwapara yuga, rapidly ascending towards the satya yuga, which represents our closest approach to Sirius.

Could it be the closer we get to Sirius the more illumined we become in an evolutionary sense as well? Is this why so many cultures around the world have followed the path of Sirius with such intense interest?

CHAPTER 22
MUTATIONS AND NEW
SPECIES

Just as the heliopause protects our Solar System from interstellar dust, cosmic rays, and gamma rays, so the Earth's magnetic field protects us from radioactive particles generated through solar winds, cosmic rays and gamma rays. During a magnetic reversal we lose this protection, and cosmic radioactivity inundates the planet. This radioactivity could be 2,000 times greater than normal, resulting in massive mutations on a scale rarely encountered under normal conditions.

In his book, *Magnetic Reversals and Evolutionary Leaps: The True Origin of Species*, Robert Felix associates magnetic reversals not only with extinctions but also with the creation of new species. Douglas Vogt, in his book, *God's Day of Judgment: The Real Cause of Global Warming*, concurs.[1]

Although they differ slightly in their dating of events, Vogt and Felix both see magnetic reversals as the mechanism that drives evolution on Earth. Evolution is not gradual, as Darwin believed, but happens in bursts, instigated by magnetic reversals during which cosmic radioactivity is at a peak.

English biologist Thomas Huxley was a close friend of Charles Darwin. While he supported Darwin's theory of evolution, he did not believe it was necessarily gradual. Many paleontologists such as Stephen Jay Gould agree. *"Gradualism is not a fact of nature,"* Gould says. *"Most new species appear with a bang... Fossil records demonstrate that a species remains unchanged for millions of years before abruptly disappearing, only to be replaced just as rapidly with a species that is, though clearly related, substantially different. Nature does take leaps."*

Randall Carlson goes into vast detail about how and why ideas of gradualism ended up replacing catastrophism in the late nineteenth and early twentieth centuries, based on the assumption that the present is the key to the past, and how this outdated idea is now being rapidly reversed. In other words, geologists have erroneously assumed that the timing of events in the past always followed the same scale of geological change as we experience in our current Holocene geological age, which is a very thin slice of Earth's history.

The past is the key to the future, claims Carlson instead. Developing a true understanding of past geological events allows us to predict and understand long-term climate changes, catastrophic Earth changes, galactic cycles, and even to predict and understand cycles of human presence on Earth. He speaks about massive catastrophic events that shook the Earth 12,900 years ago as we moved out of the previous ice age into our current interglacial period, and how similar catastrophic changes and extinctions could shake the Earth again as we prepare to enter the next ice age.[2]

Mammoths, sloth bears, and numerous species of large mammals became extinct 11,500 to 12,000 years ago. Neanderthal man went extinct 34,000 years ago. But as we have seen, magnetic reversals release huge amounts of cosmic radioactivity, and new species have been just as suddenly birthed during these times. Our own Homo sapiens species could have entered simultaneously with the departure of the Neanderthal. Evolution is not about survival of the fittest, jokes Robert Felix, but arrival of the fittest.

It is worth mentioning here that Neanderthal man, contrary to popular belief, may have been superior to our own species in many respects. It is they, according to researchers like Colin Wilson, with their elongated heads and active pineal glands, who might have been responsible for the feats of pyramid building across the ancient world.[3]

Carlson believes that the shift from the last ice age to our current interglacial period, and the extinction events that accompanied this, was initiated by a meteorite impact. Felix discards the meteorite theory, claiming instead that these same catastrophic events were triggered by the collapse and reversal of the Earth's magnetic polarity.[4]

But perhaps these two events are linked together through galactic superwaves. As highly charged cosmic and gamma rays from the galactic center break past the heliopause, pushing immense amounts of cosmic dust and debris inside our Solar System, our Sun goes into a highly active phase, releasing massive solar flares and triggering a magnetic reversal on Earth. Simultaneously, stray comets and meteorites break past the gravitational defenses of the outer planets, and impact the Earth, leading to geological upheavals of all kinds.

These events have happened before, and will happen again, both researchers agree. Long-term solar cycles, magnetic reversals on Earth, and the abrupt climate changes that generally follow, seem to both initiate and terminate ice ages.

Meanwhile, the Earth magnetic field has been weakening steadily for the past few hundred years. In recent years this decline has accelerated significantly. Studies show that this kind of accelerated decline generally precedes a reversal, which takes place once the field strength is down to approximately 15 to 20%.

A University of Berkeley study in 2015 revealed that magnetic reversals happen suddenly, and could begin and end in the space of a single human lifetime.4 In January 2011, the British Geological Survey stated that a magnetic reversal was imminent. Climate researchers Ben Davidson, in his thoroughly researched website, Magneticreversal.org, as well as YouTube channel, Suspicious Observers, shows very graphically how we might be in the very last stages of decline before the reversal.[5]

The magnetic field strength dropped about 10% between the early 1800's and 2000, at the rate of roughly 5% per century. It fell another 5% between 2000 and 2010, and yet another 5% between 2010 and 2015; and continues to decline exponentially. The

European Space Agency, which on November 22, 2013, sent up an array of three satellites to specifically measure these magnetic fields, corroborates these findings. This array, known as SWARM, indicates that the magnetic fields are weakening 10 times faster than previously realized.

Vogt and Sultan state in *Reality Revealed* that, for an evolved soul, the time during a magnetic reversal is the only time in which to live. This is where the potential exists to evolve into the next higher dimension. This is where we can consciously mutate our bodies and our consciousness to entirely new levels of divinity.

I believe this is what happened during the final days of Atlantis in the previous cycle of magnetic reversal. Those who were prepared for it could move on into an ascended state. Those unprepared sank beneath the waves into the cycle of reincarnation, gradually evolving to a point where the same souls (in different bodies) are given another opportunity to make the leap. Ideally we can make the leap collectively as a human species this time around.

The 12,000-year galactic pulse seems to me, the Grand Initiator of all these cycles. Galactic superwaves stream out from the center of our galaxy, carrying the evolutionary potential for the next cycle of creation. As this superwave moves through our Solar System, it triggers cyclical events within the Sun, which in turn triggers a geomagnetic reversal. This leads to climate changes, Earth changes, mutations, and evolutionary leaps in response to a deep creative impulse within the heart of matter.

As far back as 2011, it was being reported that the North Atlantic portion of the Great Conveyor Belt had come close to halting. As a result of this, the atmospheric jet streams previously linked with these undersea loop currents are dissipating. Instead of bringing warm air up to regions in Europe and North American, they are causing severe storms and flooding in some parts of the world and bitterly cold winters elsewhere. The trend is likely to continue.

Dr. Gianluigi Zangari of the Institute of Atmospheric Sciences and Climate in Italy goes on record to say that, with the shutting down of the North Atlantic thermohaline current, a mini ice age has already begun.[6]

I wish to reiterate that a Little Ice Age, which follows a cycle of 206 years, is different from a Classical Ice Age, which follows an 11,500 or 12,000-year cycle. It is important to recognize that these events are part of a natural cycle. Some scientists have blamed BP for the hundreds of millions of gallons of oil made to sink by dispersants such as Corexit. These dispersants have entered the Gulf Stream and have apparently clogged it up. Although there may be truth in this, it seems the thermohaline loop had already begun to slow down prior to this. It might have shut down anyway, sooner or later.[7]

More importantly, what happens next? Would our entry into a Little Ice Age trigger the advent of a regular Ice Age? Will this be accompanied by a magnetic reversal? Is this the beginning of a global domino effect?

I feel some shock in the recognition that the climate changes that were once only a frightening possibility have now become devastatingly real. We cannot go back now. So what lies ahead and what choices remain?

CHAPTER 23
EXTINCTION SCENARIOS

I remember when I first came across the work of Dr. Guy McPherson, as cited in the chapter on global warming. Listening to his doomsday predictions, my mind filled with statistics and observations of how humans have become a cancer on the face of the Earth, and with the immense burden I had so long carried for being able to somehow contribute to changing this, I felt a deep jolt of shock and sadness. But paradoxically, I also felt a certain sense of relief.

As long as I felt there was a chance we could turn things around, there was still hope, and this hope created a burden of responsibility, which brought with it its own anxiety and stress. As long as I felt there was even a slim chance to change something then the fear of failure loomed ever more oppressively over my head.

It is perhaps similar to a seriously ill patient who gets a verdict of impending death from the doctor. Although there may be shock and despair, there is also masked within those emotions a sense of relief. He can stop fighting, and take a deep breath knowing that he is finally, for the first time in his life, present to this moment. For it is often only when confronted with our own mortality that we are faced with the truth of who we are, which brings us into the glorious inescapable state of presence.

So now here was someone telling me that it is too late to change anything, that hope is a bad idea, and given that we cannot change anything anymore, we might as well enjoy the preciousness of the time we are given, learn to love, fill our life with meaningful pursuits, and practice the art of living before we die.

McPherson's prognosis is based on his belief in the global warming scenario. Meanwhile, John Casey and others point to equally dire scenarios based on a global cooling scenario. Either way, we could be well on our way to what geologists refer to as an Extinction Level Event.

It is interesting to look at Douglas Vogt's dating for a magnetic reversal in 2046 AD. Meanwhile, Casey claims that we have already entered a time of solar hibernation that would last 30 years. This would end in 2046, precisely when Vogt predicts huge coronal mass ejections from the Sun initiating a magnetic reversal on Earth, accompanied by volcanic activity and earthquakes. Is this the moment when the next major galactic superwave would pass through our solar system?

There have been enough apocalyptic dates thrown around in recent years that I tend not to pay undue attention to them. But still, it may be wise to pay heed to cycles of history and time as we prepare for what might lie ahead.

Humans tend to thrive in a very narrow band of inter-dependent factors, and it takes very little to throw us into terminal chaos and disarray. During the mini ice age of the 17th century, extreme cold temperatures initiated crop failures, famines, riots, and war, sweeping through much of Europe, and decimating a fair percentage of the population. What effects are we likely to experience if this happened again on a planet with 7 billion people?

Or what if we experienced a Carrington event, such as in 1859 AD, when an electromagnetic spike from the Sun destroyed global communication systems? The only communication systems that existed in those times were telegraphic networks. Imagine the chaos if something similar took place today in a civilization that has become so highly dependent on computers, power lines, mobile phones, kindle readers, internet clouds and satellite networks!

The Earth's magnetic field serves as a shield against coronal mass ejections and solar flares emanating from the Sun. But what happens as we get close to a magnetic reversal and this shield begins to weaken exponentially? Even a relatively moderate electromagnetic spike from the Sun during this time could completely destroy these networks, signaling a rapid collapse of civilization as we know it.

I do believe we need to prepare for these scenarios by having enough food, water, shelter and basic resources to survive at least a year. But I also believe we have choices based on evolutionary programs that are being activated within us now, choices capable of shifting timelines and altering external realities.

I have hope in the reality of a living planet who has her own evolutionary drive and destiny, and who may have some compensatory surprises up her sleeve. It may be that the short-term cycle of global warming has somewhat mitigated the longer term natural cycle of global cooling, or that methane released from Arctic ice in the wake of underwater volcanism could help compensate against the extreme effects of a solar cooling cycle. Gaia is an open, conscious, living system, and She may be more resilient than we imagine.

I also have hope in the knowing that we are vast divine beings with limitless potential, learning now to unlock ourselves from the matrix of fixed external outcomes. I have stopped placing my hope in institutions, nations or governments. But if even a few of us can learn to shift our perceptions and biology towards the genetic requirements for a new species, perhaps we may still transform things on Earth, however dire they currently seem.

Yes, Homo sapiens as a species may be on its way out, as McPherson and others clearly state, but perhaps all this is simply setting the stage for the next species, Homo luminous, to enter.

Hope may be a bad idea, as McPherson reminds us, but it can also be a gift. False hope forces us towards deeper denial and fear, while true hope inspires us to shed our personal egos, align with something bigger than ourselves, and do what is required to see a brave new world birthing itself from the shadows of smog filled cities, dead oceans, and environmental wastelands.

What constitutes the source of our hope is the question we must ask ourselves now. These scenarios, however pessimistic they may seem, provide for humanity a clear call to consciously step forward into our highest evolutionary destiny. For it is only then that we have any chance at all to survive and thrive.

If the challenges facing us are strong, the spirit driving us from within is even stronger. We are facing the same choices between extinction and mutation that our ancestors faced so many times before. Yet it is important to remember that we are not simply victims of fate. We are 'wide-browed creators' awakening within the dance of new creation. Our choices are infinite.

What new worlds could potentially emerge from within this dance of creation? Albert Einstein reminds us that we cannot solve the immense problems that stand before us from the same level of consciousness that created them. On the other hand, what possibilities can we dream awake by entering into a supramental realm of human consciousness?

The following section of this book offers some perspectives for understanding and implementing this wider hope.

PART III

A PLANETARY BIRTHING

There is a river flowing now very fast. It is so great and swift that there are those who will be afraid. They will try to hold on to the shore, they will feel they are being torn apart and will suffer greatly. Know that the river has its destination. The elders say we must let go of the shore, push off into the middle of the river, keep our eyes open and our heads above water. At this time in history, we are to take nothing personally, least of all ourselves. For the moment we do, our spiritual growth and journey come to a halt. The time of the lone wolf is over. Gather yourselves. Banish the word struggle from your attitude and vocabulary. All that we do now must be done in a sacred manner and in celebration.
We are the ones we have been waiting for.

Message from Hopi Elders, 2001

CHAPTER 24
THE IMPOSSIBLE DREAM

My father was a scientist, and my mother an artist. Although I have been on a quest for spiritual awakening for as long as I can remember, their influence in my life ensured that my quest would be both grounded in the world of material realities as well as open to the flow of infinite possibilities.

It has not always been easy. Born under the sign of Pisces, I have always been a dreamer and felt the power of those dreams. More difficult was to learn how to discern between dreams and reality, and not to lose touch with the Earth while exploring the frontiers of the cosmos.

Ultimately I learned that we are given our dreams to make them real. It is not enough to just keep dreaming. As we commit to live our dreams out in the mirrors of daily life, we discover who we are — our limitations as well as our strengths, our humanity as well as our divinity. As we embrace both, we are no longer identified with either, and a power greater than we can imagine takes over. Then the dream, freed from itself, manifests in outer form.

I once explored drama as a major in college and took part in several plays and musicals. One of my favorites was Man of La Mancha, a musical portraying the well-known story of Don Quixote and his faithful squire, Sancho Panza.

The Impossible Dream

Don Quixote viewed the world differently from most people. He rode a tired old horse that in his own eyes was a magnificent charger. He strung together bits and pieces of tin, which became for him a suit of shining armor. He tilted at windmills thinking they were giants, and he believed in chivalry during an age in which kindness and honor were lost.

Yet for all that, his madness touched peoples' hearts, and showed them a hope that transformed their lives. He refused to believe in a flat, insipid world of artificial realities. And as he believed, so it was. His 'madness' became the door to a greater sanity.

The theme song remained with me long afterwards. I see Don Quixote in my mind's eye now as he rides off through the mist, followed by his ever-faithful Sancho Panza, *"To dream the impossible dream, to fight the unbeatable foe, to bear with unbearable sorrow, to run where the brave dare not go."*

I realize that our own world is not very different sometimes. It isn't always easy to live in a 'real' world of facts and values that often seem to be so deadening and unnatural. Eventually I learned that there is no reality separate from what we create within ourselves, and that the outer world is simply an extension of our own deep experience of truth.

For readers familiar with astrology, my ascendant in Sagittarius is exactly conjunct the position of the galactic center, which may explain why I have always felt so attuned to a galactic heartbeat of alternate possibilities. Interestingly, Pluto has crossed the galactic center in recent years, and is now going through Capricornian realms of material density, dissolving structures and systems that no longer serve. This may explain why the theme of personal and planetary rebirth has accrued so much significance in these times.

The USA, as a nation, has in many ways represented the future of humanity, mirroring the very best as well as worst aspects of our collective psyche. Founded as a republic in 1776, the US now faces in 2024 its first Pluto return, a time of facing the deep shadows of genocide, ecocide, violence, racism and greed, and transforming this into a blueprint for a sustainable breathable planet where all voices, all cultures, and all species of life are honored and welcomed.

Can we rise to this task? Perhaps we all need a little bit of Don Quixote as we enter into this next section of the book.

CHAPTER 25
A GALACTIC HEARTBEAT

Our galaxy has a heartbeat. LaViolette's galactic superwave is the physical representation of this, implying that the galactic core pulses in a mostly steady rhythm of 12,000 years for each heartbeat.

This 12,000-year heartbeat is a rough estimate based on my current attempts to correlate various aspects of scientific research with Yukteswar's mystical understandings. LaViolette proposes that specific pulses could vary a few thousand years from this average and do not occur with clockwork precision.

It is possible that more precise dating of the ice cores will improve our knowledge about this periodicity, especially as we better understand the relationships between what has been perceived until now as unconnected phenomena, including galactic superwaves, magnetic reversals, cosmic mutations, ice ages and evolutionary leaps.

The Maya understood the spiritual dimensions of this heartbeat as well, and tried to attune themselves with it. Their Tzolkin calendar represents 260 facets of this galactic energy, as externalized through our human psyches.

A calendar system represents an orientation in time and space. A solar calendar representing our 365-day journey around the Sun keeps us anchored within a third-dimensional context of space and time. The galactic calendar of the Maya provides us with a bigger picture, anchored within a higher-dimensional galactic pulse of space and time.

Some physicists, including Vogt, Sultan, and LaViolette, are coming to believe that the center of our galaxy, like all galaxies, derives its energy from outside of the space-time continuum as we know it. If the Big Bang theory is accurate, then there should be a single location in the universe from which everything expanded outwards. Instead, astronomers are finding that wherever they look, the universe seems to be expanding outwards at the same rate.

Vogt and Sultan perceive this to mean that our physical universe is a reflection of a dimension beyond the visible universe, which they refer to as the diehold. Matter and energy are continuously spilling into our physical dimension from this diehold, as their book explains.

Similarly, LaViolette's theory of continuous creation posits that the universe did not simply begin with a big bang, nor will it end with one, but is continually birthing itself. New creation is happening all the time. We can see this most clearly in the centers of galaxies.

LaViolette's theory of sub quantum kinetics implies that there is a sub-space dimension beyond the world of physical creation from which physical universes are born. This higher-order dimension underlies the quantum level, which he refers to as the transformation dimension. Thus the center of our galaxy itself, he theorizes, is not a black hole as many astronomers claim, but rather a Mother Star through which new creation is continually being birthed.

What kind of birthing are we talking about here? Are we talking about stars, dimensions, forms, consciousness… perhaps all of the above? Is the galactic superwave not just a physical burst of cosmic ray activity emanating from the center of our Milky Way galaxy, but a new wave of creation carrying with it infinite streams of evolutionary possibility?

I believe that we are multidimensional beings and that we experience different realities on different levels of existence. Aspects of our being exist beyond the boundaries of space and time, unified with galactic consciousness. From this level of existence we can influence and direct physical reality within space and time. If we can expand our consciousness to link with galactic consciousness, we can determine the nature of the new energies coming in and what they can be used for. We are not separate from these energies and can therefore choose to direct their power in whichever ways we choose.

I have come to believe that the galactic superwave is not simply a stream of mindless cosmic rays flowing out from a gigantic black hole, but is rather packets of information emanating from the Mother Star that can be programmed to create certain effects upon reaching us in space and time. It is like the ancient Indian metaphor of Indra's net in which the universe is likened to a vast net of pearls spread across the Sky. Each pearl is constantly reflecting and being reflected within every other pearl. Each one is instantly changing and being changed by every other pearl.

As in Indra's net, our attunement to galactic consciousness in the depths of our own being immediately invites a response from the galactic center regarding the nature and intensity of the incoming galactic superwave.

Just as neurons respond to our thoughts and bring about an instant response within our physical bodies, so our collective thoughts and attitudes invite a corresponding response from the nerve center of our galactic bodies, such as the Mayan concept of Hunab Ku, the consciousness that ensouls our galaxy.

I realize that these concepts may seem a little far out for many people. Where does this quixotic belief stem from? I would like to go back a few years to begin responding to this.

CHAPTER 26
DOLPHINS, WHALES AND GAIA

E arlier in this book I shared my experience with Windrider during the solar eclipse of 1999 and the timeline split of New Year's Eve.

Earlier in that same year I had been living on the Big Island of Hawaii in a little cabin near Kealakekua Bay — a dolphin sanctuary where wild spinner dolphins would often swim into the bay to rest and play among the shallow waters.

Each morning I would don my snorkel gear and jump into the beautiful, tranquil bay. When fortunate enough, I would meet up with the dolphins. Over a period of months we became well acquainted.

I learned to communicate with them through pictures and a language of feeling. When swimming with them, any sense of personal identity would easily dissolve. I joined with them in a place neither human nor dolphin, a place of pure joyful-playful-being.

I learned a lot about myself. Over the years I had built a whole set of identities around myself — I was a spiritual teacher and a healer, I was sensitive and compassionate, I was a good person with many gifts to offer, I was wise and deep…

The problem with all this was that I had become so identified with this self-image that these very identities became a mask. I then found myself carefully protecting this image lest someone see through me into a place that was vulnerable or uncertain, angry or unloving, fearful, depressed, or shy.

Forever comparing myself to others, my sense of self came from how I imagined others perceived me, and whether I thought I was good enough or lovable enough. And so, of course, I had to wear my best face at all times.

I was losing my sense of spontaneity, my childlike wonder. I was losing my ability to live from the heart.

The dolphins didn't seem to care for any of that. They became my mirrors. When I got lost behind my masks, whether in self-importance or in self-deprecation, they would stay away. When I let go of my masks, we would enter together into ecstatic play. Eventually, I learned that it was safe to even let go of the mask of being human. Here, we could meet in a space of pure essence, and that is when they truly welcomed me into their pod.

I will never forget the day that the four elders of the pod first swam with me cheek to cheek for a full hour, engaged in deep eye contact the entire time. It felt like an initiation into pod consciousness. I felt myself falling deeply in love with these beautiful wild dolphins, falling deeply in love with all of myself, and deeply in love with love itself. I learned about communicating from the heart. As I would enter into a state of love, and join it with an intention or a picture, they would respond immediately. I felt awed and so deeply grateful for my new friends.

I learned about my attachments as well. I loved my dolphin family and often found myself feeling more connected to their world than to my own. I would fall into bouts of great depression if a day went by and they didn't show up, and so of course they stayed away. When they did come in, it appeared they totally ignored me while playing with everyone else.

I finally realized they were illuminating my deepest fears of separation, abandonment, jealousy, and loss — so much a part of our average human neurosis. I never experienced any kind of judgment from them, but always a direct reflective feedback to whatever form of energy I emitted. I learned more from these unconditionally loving and compassionate teachers than I ever had in any ashram or meditation retreat.

Then one day I had an encounter with a mother and baby whale. Again my perception of life changed forever. As I approached the bay for my customary swim I noticed a spout of water. My heart leaped within me. Although I had seen whales often on their migration paths through those waters, I had not seen them so close to shore. I swam out in the general direction where I'd seen the spout, then closed my eyes and called out to them in the way I had learned to do with dolphins.

I opened my eyes to see a huge humpback female directly beneath me. At first I panicked. What if she chose to surface just now? What if I got flipped around or sucked under by her immense size? Then I realized that she was very aware of my frail presence, that each movement she made was very deliberate, and that she would never let harm come to me in any way. I relaxed, and we entered into a space of communion while her baby playfully wove his way in and out, through and around us.

Suddenly, I found I was no longer locked into my physical body. As I merged with the consciousness of the whale, I found myself expanding far out into the Earth, entering into her consciousness, and becoming one with her body. I was the Earth. I was Gaia, Goddess of the Earth. I was Ra, the Solar Logos. I was Hunab Ku, the Galactic Mother. I could feel the vastness of her form and of her being, and it was inside me as I was inside her. The cells of my body were the stars in the Sky.

I realized there was literally no separation between us. This was the consciousness that the whales live in all the time. They are the guardians and reflections of the very life of Mother Earth and of our entire holographic universe in a way that most of us in human skin cannot begin to imagine. I felt immensely privileged to experience this glimpse of truth.

With this realization, ecstasy filled my body, in wave upon wave of understanding and joy. There has always and forever been only one of us here! I understood, with Chief Seattle, that we have been forever connected, like blood that unites one family.

Later, as I went back to my cabin and began to write of my experience, I knew that this connection with my humpback friend would always remain, and through her, my connection with the soul of Earth and Sky. In that merging of our consciousness, a pathway was created within me that has remained open. The pathway has likely always existed, but the whale taught me to access it and to trust what I was feeling.

CHAPTER 27
SRI AUROBINDO AND
HUMAN EVOLUTION

M y encounter with the whales opened up a doorway for me to experience galactic consciousness. I have since come to believe that each galactic pulse has both a physical aspect and a spiritual aspect. Although we cannot control events on a physical level, we can certainly co-create alternative realities as we connect with the consciousness of this galactic pulse.

Dmitriev's observations seem to indicate that some form of a cosmic ray or gamma ray burst may already be passing through our Solar System. LaViolette tells us that a major galactic superwave may be on its way. In the previous section we examined some of the possible physical effects of the galactic heartbeat, focused through the galactic superwave phenomenon. In the next few chapters I would like to examine the spiritual aspects of this galactic heartbeat. I would like to begin with the work of two people whom I consider to be among the greatest visionaries of our times.

Near the beginning of the twentieth century the great Indian freedom fighter and yogi-sage Sri Aurobindo began to express a truth that had not been expressed before. In his high states of divine union, he saw that the time had come for a new stage in the evolution of mankind. He saw that the divine was to manifest right here on Earth and that the time for this divine emergence into Earth life was now. He spoke of heaven descending to Earth, even as Earth experienced a breakdown due to the intrinsic resistance held within her material body towards this descent.

A French mystic, Mirra Alfassa, who later became known as the Mother, joined Sri Aurobindo in Pondicherry, India. Together they embarked on a journey of intensive cellular and collective transformation that is very relevant to the colossal uncertainties we face today.

Sri Aurobindo saw that the divine force permeates all matter and that all matter therefore has a force of consciousness. The process of the divine spirit descending down into matter is called involution. The process by which the divine ascends upwards out of matter is called evolution.

According to Sri Aurobindo, humanity has reached a stage in which these two events are occurring simultaneously. He foresaw that we would soon experience a descent of what he called supramental consciousness that would entirely change everything on Earth. The term supramental refers to a unified state of being from a level of consciousness that has not yet been experienced on Earth.

Does this supramental descent relate to the evolutionary effects of the next incoming galactic superwave?

"Man's greatness is not in what he is," said Sri Aurobindo, *"but in what he makes possible. An immortal soul is somewhere within him, even if seldom active in most people, while an eternal spirit overshadows him, even if obstructed from descent by the hard lid of his constructed personality. There is a great divine plan in motion, even if the evidence of the outer senses seems to contradict it."*[1]

Our evolution is far from complete. As Sri Aurobindo states, *"Man is a transitional being; he is not final. The step from man to superman is the next approaching achievement in Earth's evolution. There lies our destiny and the liberating key to our troubled human existence – inevitable because it is at once the intention of the inner spirit and the logic of Nature's process."*[2]

Evolution leaped far forward with the origins of Life, and later with the emergence of Mind. According to Sri Aurobindo, we are now at the threshold of another leap, even more momentous, as we prepare for the emergence of the Supermind. He speaks about four stages in human evolution.

The first of these, briefly, is the animal human, which includes most of humanity today — rational beings possessing an individualized soul, but largely obstructed from contacting and merging with it. Our behaviors stem primarily from our instincts. Biological survival, along with the propagation of the species, is the primary goal.

Next is the human human, a species that is spontaneously coming to birth under the influence of an ever-expanding supramental field, a being that is starting to merge and live from soul-consciousness rather than mind-consciousness. Anyone on a quest for deeper meaning and wholeness, including those of you reading these words, is probably at this stage of evolution.

The third stage is the divine human, a stage we are collectively moving towards in this time of global transformation. We will still inhabit a material physical body, but will be in possession of the Mind of Light, supramental consciousness that has just begun to merge with cellular consciousness. The crystal children being born today are the precursors of this root race.

Finally comes the supramental human, our ultimate goal as a species — divinity incarnated fully within the realms of matter. The unity of matter and spirit is to be fully realized upon Earth. The higher-dimensional light body would be merged into the physical body, and the body itself would be raised in frequency to what Sri Aurobindo referred to as true matter.

This subtle physical substance, this true matter, was at the same time much more concrete than the physical world, more real, more complete, and much more powerful than the physical we are familiar with. It existed independent of physical matter, yet permeated all matter.

Within the field of true matter, penned the Mother in her diaries, the Supramental Force *"would progressively be able to express unity in diversity instead of division and limitation, truth instead of falsehood, freedom instead of tyranny, goodwill instead of jealousy, love instead of hatred, and immortality instead of death."*[3]

We would transcend limitations of time, space, and matter, incarnating all the vast possibilities of our souls. We would travel instantaneously in both space and time. We would shape shift through forms and dimensions as we choose. We would express the love, wisdom, and power of the Creator in bodies of Creation. Are ascended masters such as Thoth, Jesus, Babaji, and St. Germaine the precursors of this root race?

The great work of Sri Aurobindo and the Mother was in bringing supramental consciousness down into the cellular consciousness of their own bodies. The Mother states that the supramental realms were first unified into her cellular body on February 29, 1956. The Dream that is already present in the Supramental worlds was seeded into material consciousness at that time.

Since all matter is vibrationally connected, this meant that the supramental consciousness was simultaneously brought into the morphogenetic fields of the Earth, awaiting the moment when it would flower into full manifestation in the collective field of matter.

From the perspective of science, one effect of an incoming galactic superwave would be the bombardment of our solar system with cosmic dust. As the protective magnetic shield around our solar system is breached by the massive electro-charge of incoming cosmic and gamma ray particles, it would allow cosmic dust, ice comets and other debris normally held outside the Solar System to enter and be gravitationally attracted to our Sun as well as the larger planets.

As a consequence, the Sun would eventually undergo a radical shift in its base harmonic frequency, which would instantly ripple outwards, resulting in an increase in the base harmonic frequency of all matter within our Solar System. Could it be that the true matter that Sri Aurobindo refers to is somehow related to this quantum jump in the frequency of matter, quickening our bodies so we are more receptive to the frequencies of our souls?

In his book *Awakening to Zero Point: The Collective Initiation*, Gregg Braden also refers to a Shift of the Ages. He highlights two trends that are simultaneously happening right now, which would at some point intersect, contributing to this dimensional shift.

First, there is the gradual weakening of the Earth's magnetic fields, which we discussed earlier. Second, some measurements seem to portray a gradual increase in the higher end Schumann harmonic frequencies. Although the Schumann resonance is based on the distance between the surface of the Earth and the ionosphere, and is therefore incapable of increasing, as some people mistakenly claim, Braden is referring to natural harmonics within this base frequency. Does this perhaps represent a quantum shift into the field of true matter?

In his epic poem Savitri, Sri Aurobindo points toward what is yet to come:

> *A few shall see what none yet understand*
> *God shall grow up while the wise men talk and sleep*
> *For man shall not know the coming till its hour*
> *And belief shall be not till the work is done.*
> *The world is preparing for a big change*
> *And the responsibility to bring this about lies with us.*
> *Whether we like it or not:*
> *The frontiers of ignorance shall recede,*
> *More and more souls shall enter into light ...*
> *A divine force shall flow through tissue and cell*
> *And take charge of breath and speech and act*
> *And all the thoughts shall be a glow of suns*
> *And every feeling a celestial thrill...*
> *Nature shall live to manifest the secrets of God,*
> *Spirit shall take up the human play,*
> *And earthly life become the life divine.*

CHAPTER 28
SUPRAMENTAL
CATASTROPHE

Whenever there is a descent there is corresponding resistance. Decades ago, Sri Aurobindo recognized that in order to evolve further, humankind must first go through an evolutionary crisis, a supramental catastrophe. A stage has been reached in which the human mind has made tremendous strides on the one hand, but is left hopelessly inadequate to prevent its own extinction as a species.

He wrote, *"Humanity has arrived at a certain stage of general tension — tension in effort, tension in action, tension in everyday life — and an over-activity so excessive, a restlessness so widespread, that the whole human race seems to have reached a point where either one has to break through a resistance and rise into a new consciousness or fall back into an abyss of obscurity and inertia."*

It is time for humanity to come of age. The supramental catastrophe is the outward manifestation of this inner process. As the Mother testified before her death in 1973, a consciousness higher than the human rational mind, the Supermind, was established in the morphogenetic fields of the Earth in 1956. Our task now is to bring it down into cellular consciousness — into the daily activities of human life.

With the descent of this new vibratory consciousness, our obsolete human structures will be demolished. Sri Aurobindo predicted that during this process there will be a transition period that could become extremely chaotic as the last strongholds of greed, fear, illusion, and darkness come to the surface to be dispelled.

> *"At the very bottom of the hardest, most rigid,*
> *narrowest and most asphyxiating unconsciousness*
> *I struck upon an Almighty Spring that cast me up*
> *Forthwith into a formless, limitless Vast*
> *Vibrating with the seeds of a New World."*
>
> Sri Aurobindo

This journey into the unconscious void represents the day of purification that has been prophesied by the Hopi, Maya, Tibetans, Hindus, Christians, as well as mystics, prophets, and scientists throughout the ages. Sri Aurobindo emphasized, however, that although the changes would be enormous they do not have to be catastrophic.

When there is strong pressure for change from the forces of truth there are likely to be catastrophes because of the resistance and clash of forces. Yet, the supramental, he maintained, holds a complete mastery of things and carries a power of harmonization that can overcome this resistance "by means other than dramatic struggle and violence."

Sri Aurobindo and the Mother believed that the world stage was set for an imminent supramental descent. As the Mother said in 1968, the supramental transformation was a certainty now. It would come about with a minimum of destruction, although this could still be "considerable" and would accelerate after the turn of the century.

With all the violence and all the possibilities of environmental catastrophe in our world today, it might be interesting to ask ourselves what further degree of destruction we might be experiencing if not for this power of harmonization! How many nuclear crises have been mysteriously defused? How many times has a brewing world war been avoided? How many catastrophic Earth Change prophecies have failed to materialize? And as the galactic superwave enters our Solar

System, how many worst case scenarios discussed in the previous section of this book could be averted or modified?

In this context, I remember a time I had been living in Berkeley, California. Soon after the 1989 earthquake in San Francisco, many psychics as well as seismologists predicted that a much bigger one was on the way. A shaman friend was seeing very detailed visions of the devastation it would cause, along with a specific date and time for the event.

As the word circulated, many performed rituals and ceremonies intending to harmonize their energies with that of the land. The night before the predicted date, many felt that something had shifted. At the exact time of the predicted earthquake came the news that the Berlin wall had come down! Somehow through the unified field of intention created around the world, the powerful explosive energies within the Earth had been harnessed for a very different purpose!

Stories like this are common. I wonder if we fully realize to what extent various masters, yogis, and ordinary people in their caves, homes, and mountaintops are able to influence world events for the benefit of a human species so vigorously bent on trying to self-exterminate.

Before his death in 1950, Sri Aurobindo gave five indications that would precede the dawning of the New Era. In brief, he disclosed that knowledge of the physical world would increase exponentially, people would start becoming more psychic, there would be a concerted effort from the dark forces to maintain control, the world would move towards increasing unification through technology, and finally, many such as Hitler would rise up with the power to influence large numbers of people.

All these indications, in both their dark and light aspects, seem to be manifesting with increasing intensity in the world today. The collapse of an old world order seems imminent. We see signs of this breakdown throughout our social, economic, and political systems. With Pluto now in Capricorn, there is an increasing demand for all systems, forms, and structures to undergo a radical transformation in greater alignment with incoming evolutionary energies.

When I confront the ecological and political realities of our modern world, and speculate darkly on what might still lie ahead, I find myself greatly comforted by spiritual researchers like Sri Aurobindo and the Mother. Their visions and understandings emerged from the direct experience of a more expanded cosmic truth.

We cannot remain in a state of denial concerning the ecological and political realities on our planet. However, in affirming the new realities in our midst, we are guided in our journey towards an emerging Age of Light. Somewhere between an ungrounded spiritual escapism based in denial and a narcissistic preoccupation with gloom and doom, we can choose to intentionally and purposefully envision a brave new world based on the deepest truth of our being. In order to do this we must consciously face our own shadows, seek a genuine hope, commit to rebirth, and open to the light of our guiding truth.

We can only go as high as we are willing to go deep. It is often in the very depths that we find the greatest light and the power and the hope to sustain us. When we are willing to see and to embrace all of our deeply human, fragmented realities the light of grace shines upon us.

As we enter a new spiral of time, I feel that we will see increasingly rapid expansions of consciousness on our planet in tandem with stronger resistance. As I connect with my deepest truth, I know that we will make the Shift, although there could also be much chaos to come. The emerging consciousness may trigger the last dance of a polarized worldview. Yet, more than simply surviving the trauma of these times, we will transform into something far nobler as a species.

The Hopis believe the Great Purification will bring wars, hunger, disease, and Earth changes. These have already begun. We have transmuted much however, and the transition is likely to be gentler than previously expected.

Through the power of the world soul, we have the potential to activate Sri Aurobindo's power of harmonization — abolishing the fragmentation and separation inherent in our current state of consciousness.

These are times of a great transformation of Spirit. Although many might choose to leave, vast numbers of people will choose to awaken into their spiritual mastery and merge with their light bodies. Psychic, healing, and telepathic gifts will awaken. We will learn to walk between the worlds.

As we move through these times, let us gaze beyond mere appearances to what is real. The pathway through the valley of fear is in embracing a larger planetary and cosmic perspective. As the ancient Hindu sages said, *"Lead us from the unreal to the real. Lead us from darkness to light. Lead us from death to immortality!"*

CHAPTER 29
SUPRAMENTAL
TRANSFORMATION

Sri Aurobindo speaks about three kinds of transformation: psychic, spiritual and supramental. Psychic transformation has to do with gaining an acquaintance with what he calls the 'psychic being', the inner spark of intelligence we each carry within our deep heart that guides our way through life. As we begin to listen to this voice within the heart we slowly get weaned away from the uncontrollable fire of our passions and enter into the silence of our inner being. It reflects as a presence that shines through our eyes, creating a space of strength and stillness within the storms of life.

As we advance in the task of psychic transformation we then enter the path of spiritual transformation. We recognize that within the stillness of the heart rests an entire universe, and that this universe is not distinct from all that exists in the vast oceans of time and space. The personal self and the universal self merge into one distinct identity in our journey of self-realization.

For most people on the spiritual path this is the ultimate goal of consciousness. However, Sri Aurobindo refers to something beyond, a third path of transformation that can only begin once we are firmly rooted in these first two stages of awareness. This is not a journey up but a journey down, not into light but into darkness. It refers to a path of self-perfection that includes the physical body and the material planes. It involves reaching deep into the physical mind, or the mind of the cells, and transforming this. It requires going deep into the Inconscient, into the densest layers of our personal and collective illusions, and illuminating them.

This illusion, or Maya as the Hindus refer to it, is hard-wired into our DNA. It is the lens through which we perceive the world around us, and through which we create the illusion of material time and space. Another term we could use for this collective illusion, used by Morpheus in that well known cult movie, is the matrix. The matrix is all around us and we do not see it. It is the program hardwired so deeply into our sensory perceptions and physical structures that we literally create our own version of the universe in space and time.

It is a universe with certain built in limits that we take for granted. It is brought into existence through the sensory perceptions of the rational mind. It is driven by the laws of entropy which means that our bodies are subject to decay and death. It is framed in linear time, which means that our existence is driven by memories and programs from the past, or fears and expectations of the future, coded deep into our DNA from our collective as well as personal history.

Laws of cause and effect drive this universe, ensuring that we are trapped in a web of karmic law and limited capacity for change. This universe is experienced through the eyes of duality, and therefore reflects this back to our senses as something that exists outside of ourselves. We seem to exist as one tiny speck within a vast infinite universe.

The supramental world has a completely different starting point. It is brought into existence through the perceptions of a unified mind, which experiences itself as the One Self that embraces all things, simultaneously within all things and beyond all things. The laws of quantum physics apply to this realm of creation. We are not embedded in time and space; rather time and space are an extension of our creative Consciousness, which unfolds in time and space as universal energy dancing in fields of matter. The entire physical universe exists as a tiny speck within the all-pervading Consciousness that I know myself to be!

The supramental transformation is about bringing this unified awareness into the structures of conditioned human consciousness. The same physical mind that is trapped in the conditioning of the past, once transformed, can act as a powerful stabilizing influence for the supramental field to permeate the structures of matter.

We do not age and die because we exist in bodies of matter. We age and die because our bodies are conditioned by an invisible field of limitation centered around an over-riding belief in duality. This belief in duality becomes the matrix that then shapes our existence from that point onwards, miring us into the quicksand of entropy, karma, and fragmented existence.

What if we are willing to challenge this state of things? Einstein equated matter with energy. Mystics equate energy with consciousness. Once we truly understand this, we will identify ourselves not as personalities built around bodies of matter but as consciousness capable of entering the heart of matter to create infinite possibilities within the evolving universe, which is simply a holographic extension of who we are.

How do we accomplish this supramental descent? As mentioned earlier, the first step is to recognize that our spiritual journey is not about leaving the body but entering it fully. It is about going into the collective programming that drives our existence and dismantling this, step by gradual step. I do not think this journey can be accomplished through the rational mind, although the mind can be useful in terms of understanding the big picture and holding an intention for making this transformation.

What I do believe can help is a shamanic understanding of the universe. In the Inka tradition of South America it was believed that there is only one field of energy that permeates the universe and is the universe. This energy is in a state of constant reciprocal flow, which they referred to as ayni. The unobstructed flow of this energy is known as sami; when it is blocked it is known as Hucha. They believed that humans are the only beings on this planet capable of blocking this flow of energy and thereby creating hucha. This creation of hucha eventually leads to the perception of duality, which has shaped and conditioned our world for the past many thousands of years.

The Inkas teach a series of simple practices for releasing hucha and re-establishing the flow of sami. When sami flows freely through our energy systems we are reconnected with the upper worlds and the lower worlds and can once again find a harmonious place within the flow of the universe.

In the Inka teachings the upper worlds are linked with more expanded levels of the mind, including what Sri Aurobindo refers to as the higher mind, illumined mind, intuitive mind, overmind and supermind. The middle world is our view of the universe created by the rational mind. The lower world includes the world of the physical mind, and the deep patterns of conditioning that drive our instincts and passions. These patterns are hard wired into our genes, and shape the long journey of psychic and spiritual evolution. As long as we remain unconscious this physical mind becomes the genetic matrix that keeps us trapped in the dense quicksand of Maya, the seemingly endless cycles of karma, density, separation and fear.

However, it is possible to transform this same physical mind, otherwise known as the cellular mind, so that it becomes a powerful ally for stabilizing and channeling the supramental forces pouring down from the upper worlds. The Inka practices can assist in this work. But it requires a deep focus and strong intention.

It requires a willingness to go deep into the darkness of the Inconscient, and an innate trust in the Psychic Being, our Inmost Self who is guiding us on this journey. It requires a knowledge that the Inconscient itself is simply a densified aspect of the same consciousness that pervades all things, and where we find our ultimate identity. As we go into darkness we must know ourselves as the light. Knowing this, it simply becomes a journey deeper into the heart of our own hidden self.

Since the same Self moves through the heart of all things, whenever we illuminate the perceived density of matter in one small aspect of the collective, we illuminate and holographically transform the whole. We participate in the evolutionary journey of Gaia herself.

This was the journey undertaken by Sri Aurobindo and the Mother. As previously mentioned, the Mother proclaimed in 1956, that the supramental field was finally able to penetrate into this realm of density, through her own body and into the body of the Earth. The supramental transformation had begun. Sixty years later, this work continues to deepen within the structures of evolving humanity. What does it look like now, and where do we take it from here?

"I was on a huge boat which was a symbolic representation of the place where people who are destined for the supramental life are trained. The boat itself was made of a supramental substance. The light was a mixture of gold and red, forming a uniform substance of a luminous orange. The atmosphere was full of joy, calm, order; everything went on regularly and in silence. On the boat the nature of objects was not the one we know on earth; for instance, clothes were not made of cloth and what looked like cloth was not manufactured: it formed a part of the body, it was made of the same substance which took different forms. Life created its own forms."

The Mother

CHAPTER 30
THE CONQUEST OF DEATH

Before his death, Sri Aurobindo spoke about the manifestation of the supramental consciousness on Earth. He said it would take about 300 years until this force would be incarnated into the body of the Earth, releasing in the process a great resistance held within the cells of our own bodies. This resistance, or Maya, has held us trapped within a powerful matrix of duality where we have separated mind from heart, spirit from body, human from divine, and lost ourselves in a cycle of insecurity, violence, despair and fear.

He succeeded in bringing the supramental force into his own body to a certain extent, and transferred this force to the Mother upon his own passing in 1950. The Mother then continued this work of supramental descent until in 1956 she announced that the pathway was open so that this unifying force could now incarnate like a seed within the collective consciousness of humanity.

We are not talking merely about a spiritual awakening for humanity but a species transformation. It is a biological shift in the morphogenetic fields of our cellular DNA that defines the next human species, moving us inevitably to a stage of human evolution as different from our current species as we are to the evolution that preceded us.

The Mother continued this work until her death in 1973. She had not actually intended that this work would stop, but had hoped to continue the work on a cellular level even beyond the appearance of death. She wished simply that her body would be protected during the weeks or months it might take to complete the process. This she had confided to Satprem who was her close confidante and chronicler for the last 23 years of her life. Unfortunately, in the political climate of the ashram at that time, Satprem was barred from visiting the Mother during those last six months of her life, and the ones who were responsible for her well-being chose to over-ride her wishes and put her body into the ground the moment she appeared to have passed away.

It was made publicly known that the work of supramental descent had been postponed, and would be continued at another time.

We must understand that this work of supramental descent is a conscious act of entering the programming of our cells, down to the DNA of our human genetics, and gradually, step by step, allowing these codes to be re-written. It has nothing to do with the rational mind, emotions or instincts, all of which are products of a past evolution. As the supramental descent takes place it pushes up to the surface all the resistance held within morphogenetic fields of the past history of the human race, all the patterns of violence, aggression, greed, insecurity and fear.

It is no simple matter to incarnate the supramental consciousness. An entirely new foundation needs to be built up from the ruins of the old. The descent is necessarily accompanied by all the symptoms of our collective human resistance and suffering as it emerges into conscious awareness in order to be released.

Thus the announcement made about the supramental descent being postponed came from this unconscious internalized resistance, an attempt to slow down the unraveling process, perhaps an attempt to ease the pain of too quick a transformation.

But this was only a surface truth. What few people are aware of is that the work did not stop with the Mother's passing. In an act of great determination and courage, Satprem came to the realization that it was his task to continue the Mother's work in his own body. Just as Sri Aurobindo had transferred his supramentalized awareness to the Mother, she managed to occultly transfer her own supramental force to Satprem, who then continued the work from 1982 right up to his death in 2007.

Satprem had recorded all his conversations with the Mother over the 20 years they had been together. She had chosen him and prepared him not simply to record her own experiences of divine materialization, but to accomplish something else he had not foreseen. These conversations were transcribed and became the basis for the 13 volumes of *The Agenda*. They also became the basis of this work being continued within his own body, as chronicled in the yet unpublished 24 volume series, *Notebooks of an Apocalypse*.[1]

During one of my visits to Auroville, India, I spoke with one of a team of people involved in the translation and publishing of these notebooks, who described to me some of the concrete aspects of this supramentalization process. About six months before the Mother left her body she had managed to program her cellular consciousness so that it did not require food in order to survive. As Satprem began his own journey of conscious cellular mutation he soon arrived at this same realization. This was eventually followed by even deeper genetic adaptations such as each cell being able to exchange prana and oxygen directly with its environment, rather than through the lungs. As the supramental light continued to press downwards its luminosity began to even transform the stem cells and skeletal structures within his body.

Two months before he died in 2007, Satprem wrote saying that the great doorway, which the ancient Vedas referred to as Mahas Patah, was now open, and that the possibility of deathlessness now existed as a distinct morphogenetic possibility within the human species.

Creating a morphogenetic field is like seeding a new possibility within a genetic pool. The nature of these fields is that once initiated it makes it possible for other people to follow. Like the story of the hundredth monkey, it is becoming easier and easier now for more and more people to accomplish the same cellular mutations until, when the moment is right, these new capabilities become manifest in the entire species.

It seems to me that Satprem has played a role just as important as Sri Aurobindo or the Mother's in taking forward this work of supramental transformation. I am greatly comforted to know this. Perhaps it is not going to take two or three hundred years after all for a supramental species to be manifested. As I study world events today it seems to me that this great work is continuing to accelerate ever more strongly. It may be only a matter of years or decades now until a divine human species breaks itself loose from the worn out shell of a dying age. Indeed, given the current realities of our physical planet, this has become an urgent necessity now.

It has been 60 years since the supramental descent of 1956. Meanwhile, we are placed in a situation where not only a global financial collapse, but also the possibility of a third and final world war seems imminent. The asuric, or dark forces, are rampant out in the world. But do these forces have a reality of their own, or are they simply a response to this supramental force as it presses upon human consciousness?

The nature of evolutionary leaps is such that they only take place when the old order has become sufficiently destabilized. The asuric forces have been divinely ordained to create this destabilization. Just as it seems that we will inevitably sink into collapse and destruction, the evolutionary spark will be ignited, and a collective event will become possible.

"The work is done," said the Mother. *"Each pulsation of Love carries the universe further in its manifestation. The supramental Manifestation is realized. You see, it's not as if this world of Truth had to be created from scratch! It is all ready, it is here, everything is here... Just a little click would be enough."*

There are those who will feel called to continue this avataric work within their own bodies until it is finally done. And when it is complete we shall rise up as sun eyed children of the new dawn, architects of immortality, striking one mighty blow upon the golden doorway to a new Earth, pioneering a new species of humanity into existence.

The Secret Lies in Matter
The supreme height
Touches the most material matter
All the splendors one can experience
By going up, by getting out,
By leaving, are nothing!
They're nothing, they don't have
That concrete reality.
They seem vague compared to HERE
That is truly why the world has been created
It's in terrestrial matter, on Earth,
That the Supreme becomes perfect.

The Mother

CHAPTER 31
GLOBAL INITIATION

The nature of most spiritual initiations is the experience of a gradual ascent of consciousness followed by a sudden shift. It is like water slowly heating up that suddenly becomes steam. When the time is right, this shift could take place in the twinkling of an eye.

In the subatomic world, electrons spin rapidly around the nucleus of an atom. When energy is added, they suddenly jump from one stable orbit to another. It is like disappearing from one orbit and then simultaneously reappearing in the next one. This phenomenon is known as a quantum leap.

Are we preparing for a quantum leap in planetary consciousness? I suggested earlier that there are both physical and spiritual aspects to the galactic superwave. As we attune to galactic consciousness, we therefore have a role in determining how we experience the physical effects. The choice is ours. Could the same wave that is capable of inflicting extreme destruction also be capable of initiating an extraordinary evolutionary awakening?

The Mayan calendar is a good example of this. Although it was created in linear time, the calendar is based on deep insights into the nature of galactic time. The Maya seemed to understand something about the nature of the galactic heartbeat, even though they may not have known about the science of cosmic rays and interstellar dust. Perhaps they were able to communicate with a galactic intelligence that shaped the structure and timing of their calendar system. It could be they sensed what to expect based on previous pulses of this galactic heartbeat.

Our experience of time is a reflection of our consciousness. Our rational minds, operating within the framework of a three-dimensional universe, are only capable of functioning in a linear time-based perception of the world — a historical progression of our lives from past to future.

In contrast, Sri Aurobindo refers to levels of the mind far beyond the rational mind including the higher mind, illumined mind, intuitive mind, overmind, and finally the supermind. Is it possible that each level of the Mind offers a different perception of time and consciousness, thus opening us to the experience of many possible timelines? Could these timelines exist simultaneously?

For instance, in shamanic states of consciousness the boundaries between past, present, and future begin to dissolve. When psychics and prophets tap into future events it is because they have expanded their minds into a higher state of consciousness where their sense of presence is amplified to include events in the future. The quality and relevance of this information, the accuracy of it even, may have much to do with the levels of the higher mind they are able to access. The predictions of Sri Aurobindo and the Mother, for instance, are made from the levels of the overmind and supermind, meaning that they were able to attune to a galactic level of consciousness, much like the creators of the Mayan calendar.

As we transcend the linear dimension of time, and enter into a multi-dimensional time zone, it may become possible to enter many dimensions of reality wherein our experience of the present may embrace two or more probabilities. Like my own experience on the eve of the new millennium, as a collective we may have the choice of multiple scenarios. That is a main theme of Windrider's message.

String theory proposes the existence of 11 dimensions, most of them seemingly 'curled up' in a dormant state. As the base harmonic frequency of matter undergoes a quantum shift, would this somehow uncurl these extra dimensions, allowing our bodies to transcend the current limitations of space and time?

In order to transcend linear time and third-dimensional density we must learn to transcend the level of consciousness based on the rational mind.

Explorations by Russian scientists into psychic phenomena have shown that lowering magnetic fields tends to enhance psychic abilities and distorts our ordinary perception of time. What happens when the magnetic fields of the Earth temporarily collapse? Is this a moment where we might step out of rational-minded linear time and enter into multidimensional consciousness?

Although hesitant to make specific date predictions, LaViolette affirms the possibility that the next major galactic superwave could arrive anytime now. If these superwaves have both a physical and spiritual aspect, this next one may be designed to open a doorway between worlds, create a shift in the base harmonic frequency of matter, and collectively push us into a vibrational state where the supramentalization of our minds and bodies can be manifested.

What is the precise trigger for this collective shift from linear time to multidimensional time?

Vogt and Sultan comment, *"It is true that the period of time before and during the cataclysm is a terrible time for man to live. The suffering and hardship must be tremendous, but there is another side to it. If an individual has evolved to the knowledge of what existence truly is, he can evolve himself to the next higher dimension, this being the fifth or even the sixth dimension. He can do this only during the time of the reversal, when vast amounts of potential are available to him. For an evolved soul, this is the only time in which to live."*[1]

Their book was written decades ago. They were referencing a magnetic reversal and its effects on humanity. Although their forecast for the planetary shift may be gloomier than my current visions, I have been sensing that this magnetic reversal is the actual mechanism for triggering the multidimensional shift and even for the formation of multiple, simultaneously existing, collective timelines.

How would this work? Our bodies are electromagnetic entities aligned with the magnetic fields of the Earth. Our memories are also based on magnetic fields held within our auras synchronized with the memory fields of the Earth. This includes our personal memories, our collective memories, and also the morphogenetic field of our human species.[2]

What would happen if the magnetic fields suddenly collapsed? Imagine a videotaped recording of an event. Now run this magnetic tape through a strong magnetic field. Wouldn't the entire recording be erased?

Most of our conscious memories are held within the mental body, composed of all the imprints and sensory input we have received from birth onwards, aggregated into a personality. Our identity is held within the electromagnetic fields of the lower physical, etheric, emotional, and mental bodies. This is who we think we are. This is the little self that could be instantly erased during this intense time of magnetic reversal.

Braden refers to a collective initiation that would take place when the collapse of the magnetic fields is coupled with the rise of higher harmonics within the Schumann resonance of the Earth. Is it possible that this, along with the effects of a galactic superwave moving through our solar system, could initiate a shift in the consciousness of matter?

Just as overtones can be created within the musical scale using a guitar string or even the human voice, there are up to eight harmonics that have been noted within the basic Schumann resonance of 7.83 Hz.[2] Some researchers, such as Dr Kathy Forti, author of Fractals of God, report that these higher harmonics are becoming more discernable at this time. Interestingly, all these harmonics tally with heart and brainwave frequencies that we are capable of experiencing.

Most people have heard of beta, alpha, theta and delta states, where the brain generates frequency ranges that provide access to alternate states of consciousness. Richard Bartlett, founder of Matrix Energetics, refers additionally to a high gamma state, vibrating beyond the normal beta range, which he refers to as the genius frequency.[3]

If Schumann harmonics are a reflection of human brain states, could it be that as more of these harmonics become perceivable so we likewise begin to access multi-dimensional consciousness, integrating delta, theta, alpha as well as high gamma into the beta range of normal waking consciousness?

As our brain waves shift from delta to theta to alpha to beta, and then even higher up to gamma frequencies as we transition from deep sleep to active waking states; could this be an indication that somehow we may be waking up in a spiritual sense, both individually and collectively?

Is it possible that the increase of solar and cosmic rays entering the Earth's atmosphere due to the weakening of its magnetic fields also contributes to the activation of higher harmonics within the Schumann resonance fields?

What does a geomagnetic reversal mean in this context? Could it be that our linear mind-based ego identity is being slowly erased with the decline of Earth's magnetic fields? And that as the magnetic field of the Earth ultimately reverses itself, a new memory field could then be activated based on the remembrance of the Self? It's like pressing a cosmic reset button. We would emerge into a soul identity that includes our incarnational memories on Earth from the perspective of unity rather than separation.

We are not far from this moment of cosmic birth. As we respond to this deep longing of our souls, we may soon experience a quantum leap in consciousness, birthing ourselves collectively as the divine human!

CHAPTER 32
THREE DAYS OF
DARKNESS

Death is the door to immortality. Darkness is the gateway to light. There are many prophecies from around the world that refer to the three days of darkness we must pass through as we transition into the new world. On a visit to Guatemala many years ago, Mayan elder Don Alejandro Cirilo Perez Oxlaj spoke about these three days of darkness as the last and final gateway in our journey towards Year Zero.

I have come to feel that these three days of darkness are referring to the time of magnetic reversal. Magnetic reversals are the key to transformation. As with the dark tunnel reported by many people who have had near-death experiences, this three-day period is the tunnel we must pass through in our journey of planetary rebirth.

Are these prophecies referring to a time of literal darkness where the Sun refuses to shine? Or are they referring to a metamorphic dark night of the soul where our personal identity eventually gives way to a soul identity? Are there elements of both?

It seems to me that just like a caterpillar going through the darkness of the cocoon in its emergence as a butterfly, we too may be undergoing some kind of metamorphosis. I can't imagine a scenario where all of Earth would be cut off from the light of the Sun unless we were simultaneously going through some kind of dimensional shift where time-space coordinates also undergo a shift.

Going through a dimensional shift would be like entering into a collective cocoon where our physical senses temporarily disconnect. This would happen not just with humans but also with all species and with the consciousness of the planet as a whole. This could well be the mechanism for evolutionary leaps that Felix equates with times of magnetic reversal.

Is this why we have almost seven billion people on Earth today — so that we all get a shot at this quantum leap in evolution? We are not just referring to spiritual evolution but biological evolution as well. We are talking about the possibility of a newly emergent species of humanity, along with innumerable others. Such a unique event becomes possible during these times of magnetic reversal once every 11,500 or 12,000 years.

It is said that when Atlantis sank, those who were energetically prepared for this were able to use the incoming energies of the superwave blast to merge with their light bodies. Those who were not able to do this, or who allowed themselves to be engulfed by fear, disappeared beneath the waves to start the entire cycle of human evolution over again.

We stand once again at this same threshold of awakening. According to Windrider, the time traveler I encountered during a shamanic journey in 1999, the probability of collectively remaining in linear consciousness during the next galactic superwave blast, which would in all probability create large-scale cataclysmic events similar to what caused the sinking of Atlantis, no longer exists.

Instead, as we pass through this collective doorway, we would find ourselves able to experience higher levels of the mind, allowing us to embody more of our souls and transmute many of the cataclysmic aspects of the global Shift.

As Don Alejandro told us, *"When you find yourselves in these three days of darkness, go inside your homes, close the doors, and celebrate. These are the times you have been waiting for."*

I believe that when we collectively move out of linear time, we will be able to consciously accelerate our own evolution. We may remain in third-dimensional reality in this or another world if we choose to continue exploring a state of polarized conflict. Or if we wish, we can choose to experience increasingly more subtle, light-filled realities in fourth, fifth, or sixth-dimensional worlds.

When I am talking about dimensions here I am referring to vibrational frequencies. The third dimension has to do with linear time and dense physical matter as we know it. The fourth is a transitional dimension that begins to shake us out from our perceptions of rigid space and linear time. The fifth and sixth dimensions provide us with experiences of multidimensional time and increasingly subtle physical substance — the realm of true matter.

Hypnotherapist Dr. Chet Snow has written a fascinating book called *Mass Dreams of the Future*. In this book, rather than regressing people into past lives, he "progresses" them into the future, beyond the time of the Great Shift, and then catalogues their experiences.

Interestingly, their experiences fall into three very distinct categories. The first group was living on a devastated Earth, foraging and living in hunting and gathering conditions similar to how it must have been after the Atlantean catastrophe when humanity had to start over from scratch. The second group either lived off-planet in space stations, or on the surface of a devastated planet in an artificially created environment. The third group lived on a beautiful green utopian Earth, completely restored. There was no crime, oppression, war, or hunger. People lived in harmony and peace with each other and with the planet.

In some of my own work as a hypnotherapist, there have been occasions when people went even further on to worlds where they experienced themselves in bodies of light. I share a couple of these transcripts later in this book.

Could these be examples of what we are calling third, fourth, fifth, and sixth-dimensional timelines? Individuals may be able to choose vastly different realities after the Shift, in accordance with their inner soul choices.

Do these dimensions roughly correspond to Sri Aurobindo's vision of future human evolution? Does the fifth dimension represent the birthing of the Divine Human, and the sixth the advent of the Supramental Being? It seems to me that the magnetic reversal would initiate our experience of these higher dimensions of reality, whether this happens simultaneously in a multiple worlds scenario or progressively in evolutionary stages. All of this would be assisted through the power of harmonization that Sri Aurobindo discussed.

If enough of us are able to ascend to this level soon enough, this power of harmonization could bring about Windrider's best-case scenario — a complete and immediate collective transformation of Earth and humanity into a frequency of matter fully capable of holding the vibrations of soul. If this works, it could trigger the hundredth monkey phenomenon, where we ascend together as a planetary collective into a higher-dimensional timeline of human consciousness!

If we are not quite ready for this collective transformation, perhaps the magnetic reversal would initiate a many worlds scenario, splitting our third-dimensional space-time coordinates into several simultaneous timelines, each one vibrating to a different dimensional frequency, with each of us given a choice of which timeline would best serve our continuing evolution.

When is this likely to happen? Most researchers that I have referred to in this book will not give specific dates for the manifestation of their respective scenarios, although the consensus is that this could be relatively soon. Given the current state of affairs, I see the upcoming global Shift as a sign of enormous hope. Many informed people around the world intuit that we cannot endure another ten, twenty, or fifty years without exterminating our precious planet and ourselves. Most have become quite pessimistic regarding our ability to turn things around in time.

The conclusions that we are making here are not about death and destruction, which would inevitably happen anyway if we continue our current course. Rather, I see these as visions of hope and beauty beyond imagining. This is especially true once we discover how much power we hold within our own hands to determine our personal and collective future.

CHAPTER 33
THE MANY FACES OF HOPE

There are times I feel a deep sadness about the choices we have made on Earth. We have wrapped ourselves in a tight net of belief systems and conditioned behaviors that are slowly destroying our collective soul. I remember in some buried part of my cells how it used to be, our relationship with nature, our relationships with each other, the sacredness of life springing alive and fresh each moment. In our insecurities and fear, resigning ourselves to the fear of death and restriction of form, we have lost the spontaneity of life.

The Sioux warriors of North America had a saying. When they gathered together around the evening fires to sing and dance with their women and children — celebrating the simple events of their lives — they would tell each other, *"It is a good day to die."* And when they rode out to battle, not knowing if they would see another sunrise on the prairie, or live to watch their children grow up, they would remind each other again, *"It is a good day to die."* Life was good, overflowing with beauty and lived fully. As they welcomed their life, so they would welcome their death.

Somehow, when I was young, I was convinced that I would not live beyond the age of 30. So I decided to live my life as fully as I was able, so that when the time came for me to die, I would have no regrets. My greatest fear was not that I might die, but that I would not have fully lived. Going towards the edge of my boundaries, I literally fell 200 feet off of a waterfall once, and survived. The experience opened me up to the spirit worlds. By the time I got to my thirtieth birthday, still incarnate in this body, I learned that it was also a good day to live!

Death and life go hand in hand. Don Juan, the Yaqui medicine man popularized by Castaneda, spoke of keeping the awareness of death ever-present at your left shoulder so as to live in total awareness of each moment. Are we willing to enter into our fears of death deeply enough to recognize our fears of change? Could it be that our fears of change are what hold us back from being fully alive?

For many of us, things are changing too quickly, and it is not easy to keep pace. Our finances, our health, our relationships, and our ability to trust ourselves and the universe, seems to be increasingly falling apart. We spend our lives looking for security. Yet in a world of increasing chaos, this is not so easy to come by. It all comes down to trust. How do we trust, whom do we trust, and why?

If this is your first time encountering some of the material in this book, I imagine that you could be experiencing a crippling combination of shock and fear. There is so much fear and uncertainty in our lives already. We have fears of making ends meet, of raising kids in an increasingly violent world, of losing someone we love, of intimacy, of being alone, of not having what we want, of getting what we don't want.

There are fears of being controlled by somebody else. We have all had religions, governments, parents, partners, or various   authorities tell us what we can or cannot do, what we should or should not feel, and what to believe. There are also fears of having to think, feel, or act for ourselves. There are fears of failure, fears that we are not good enough, or don't know enough, or don't feel enough. There are fears of our own unhealed darkness.

Ultimately, as Marianne Williamson so eloquently reminds us in her book, *Return to Love*, "*Our deepest fear is not that we are inadequate. Our deepest fear is that we are powerful beyond measure. It is our light, not our darkness, that most frightens us.*"[1]

On a global level, the realities portrayed by our news media are becoming increasingly fearful. Some of the things not covered in the mass media, as described in this book, can be even more fearful. Yet as we take the time to connect with our light, open our minds to the big picture and our souls to new possibilities, these same events can be cause for great celebration and joy!

I have spoken often in my groups about averting the Earth changes through our collective intentions and prayers. While I believe this perspective has validity, we cannot pretend to understand what evolutionary transformation must look like. There are certain cycles of death and rebirth that serve our evolutionary growth, and a massive dissolution and restructuring of our planetary realities, including some degree of economic, political, and Earth changes may accompany this next cycle of evolution.

The easiest way to make the transition is to understand it from an evolutionary perspective so we can choose to joyfully align with the process rather than resist in fear and contraction. With all the apparent realities of terrorism, war, tyranny, and catastrophe in the world today, it is still possible to live with increasing fearlessness, excitement, and joy.

My hope for the world comes from a place deep within me that understands fear and denial, yet no longer lives there. As long as we are in denial we can see neither darkness nor light, and I see that many of us are on a journey of fully embracing our fears so we can release our illusions.

Once we understand the nature of illusion, we realize that anything we perceive as a 'threat' from the outside, whether it is Earth changes, terrorism, political repression, or war, are ultimately our personal unowned shadows being reflected back.

As Ramesh Balsekar wisely reminds us, if we are not identified as the doer, there is nothing to fear. There is no blame, no guilt, no shame. All is unfolding perfectly according to a cosmic plan. We can always find something to feel guilty, fearful, or regretful about, if we so choose; or, we can simply let go of our need to control events, and decide in that moment to be free.

Paradoxically, once we are no longer identified with the doer, creation happens by itself. Magnified a hundred times over as a transcendental unitive power, creation overshadows these puny body-mind organisms and acts through our surrendered egos. The trickle of our own limited capabilities expands into a dancing river of evolutionary possibilities.

Once that understanding comes, we can choose with the enormous power of the newly emerging world soul to co-create a new unified reality. Beyond any outer appearances, even beyond any cosmic or planetary cycles, we are each responding to the supramental descent, the in-breath of God, the return of all things back into Oneness.

Ancient prophecies, along with contemporary indigenous voices, inform us that we stand at the doorway of a massive Shift of the Ages. This could mean climate changes, magnetic reversals, geographical pole shifts, shifts in consciousness, or all of the above. It could even be a dimensional shift, the end of time as we know it. Beyond this doorway, according to the Maya, nothing is known. It is a blank slate and we hold the chalk in our hands.

Meanwhile, Sri Aurobindo and the Mother tell us that a supramental transformation is imminent, and describe what this might look like. Will the human race rise up together into this experience and rebuild the Earth according to our highest collective dreams?

I saw them cross the twilight of an age,
The sun-eyed children of a marvelous dawn,
Great creators with wide brows of calm,
The massive barrier-breakers of the world,
Laborers in the quarries of the god...
The architects of immortality.
Into the fallen human sphere they came,
Faces that wore the Immortal's glory still...
Bodies made beautiful by the spirit's light
Carrying the magic word, the mystic fire.
Carrying the Dionysian cup of joy,

Gaia Luminous

Lips chanting an unknown anthem of the soul,
Feet echoing in the corridors of Time.
High priests of wisdom, sweetness, might, and bliss;
Discoverers of beauty's sunlit ways...
Their tread one day shall change the suffering Earth
And justify the light on Nature's face.

Sri Aurobindo, 'Savitri'

CHAPTER 34
THE SUPERMIND

The term evolution has rapidly become a new buzzword in scientific as well as spiritual circles. But what does it really mean?

Is it merely a Darwinian phenomenon operating through accidental permutations of slow and gradual genetic mutations through natural selection? Is it a force within matter that blindly pushes life forward into a higher spiral of expression? Is there a field of intelligence beyond matter, which is continually involving into matter as matter simultaneously evolves towards this higher intelligence?

Are we even asking the right questions?

I believe that as long as we are approaching this question from a dualistic framework where mind, matter and consciousness are held separate from each other, any response to this question will be limited and unsatisfactory. We are conditioned to experience objects in nature, Earth, universe, and everything outside of ourselves as separate from ourselves.

It is a perception of reality that we don't question. It is simply the way things are. Matter is the foundation of life and has a reality of its own, derived somehow from the Big Bang of original creation. From this matterialist perspective, where all objects and organisms have a life and reality that is fixed, dense and separate from the rest of creation, Mind slowly evolved within Matter during a long course of biological mutations, and Consciousness slowly evolved within Mind. And now somehow we have achieved a pinnacle of evolution where humans can reflect upon the nature of consciousness as masters of creation, imagining that we hold the power to direct the next stage of our own evolution.

But is this perception of reality accurate? What if we reversed this equation? What if Consciousness is primary in the field of Existence? What if Mind derives from Consciousness, and Matter derives from Mind?

Copernicus created a revolution in human thought when he theorized that it was the Earth that circled around the Sun and not the other way around. Could we be on the verge of a similar revolution in human consciousness as we assimilate the implications of such a radical change of perspective?

This perspective is not new. Mystics and sages across the ages in many parts of the world have understood this profound secret, known by the Vedic seers as advaita, or non-duality. Matter and consciousness are not separate. You and I are not separate. There is only one fundamental reality in all of existence. It is only a lens of the mind that creates the illusion of separation.

This was not merely seen as a spiritual truth, but was a practical understanding that influenced the arts, sciences and technologies of many ancient civilizations in previous cycles of time. Today it seems that our scientists are leading the way back. Quantum physics has succeeded in deconstructing the illusion of matter.

Many physicists today are becoming convinced that there are no fundamental building blocks of matter, that matter and energy both find their origins in a subquantum kinetic vacuum that goes beyond our current understandings of time space and creation itself. The properties of this kinetic vacuum are the same properties that the early Indian sages attributed to the primal field that exists beyond time space and creation, the field of Consciousness that holds all things as potential within Itself, known to the sages as Sat-Chit-Ananda.

This primal universal field, reflecting within and beyond all things, is that from which all things derive their existence. Like a deep ocean creating ripples and waves upon its own surface, it manifests itself as ripples of matter and energy that are constantly appearing and disappearing in a cosmic dance of creation and dissolution. The physical universe emerges from within this dance like a string of pearls, where it is said that each pearl reflects every other pearl, contains every other pearl, and is every other pearl.

The medium through which this dance of creation takes place is the Mind, sometimes known as Ishwara or Maya, of which the human mind is but a fractional and holographic reflection.

So we have here the vast ocean of Consciousness, of which the visible known universe is a tiny fraction; we have ripples of Matter along the surface of this ocean, manifesting as endless constellations of stars, galaxies, atoms and organisms; and we have the lens of Mind which allows us to observe this distinction and participate within this grand illusion of creation as if it were real in itself.

As we observed earlier, the nature of Mind as it currently functions within the human organism is inherently dualistic, but nevertheless extremely powerful. Within this dualistic perception we have conditioned ourselves to believe that our existence as separate individuals is real in itself, that evolution shapes itself through blind chance, and that all of history is attached to a linear timeline. As we believe, so it is. The power of rational mind has created a matrix of reality that we believe to be real, and which now pervades the entire fabric of human history.

As we continue to act out our belief in a separate existence, we have come close to annihilating ourselves and our planetary ecosystem in the dark dance of greed, insecurity and fear that inevitably result from this mindset. If this quality of Mind were the only tool we had to perceive and create external reality we are surely lost as a planetary culture. Whether we go out in a blaze of nuclear glory, or a whimper of chemical asphyxiation, is not important. Evolution has reached a point of stalemate where we cannot turn the clock back; yet remain unable to solve our mushrooming problems as we move ahead.

But what if there is a new evolution that can now begin to express itself based on an entirely different quality of Mind? If we go back to the model of reality where Consciousness is primary and Material Reality is created as the product of the Mind, would our experience of reality change according to what quality of Mind we are capable of manifesting?

As experienced in his own states of yogic contemplation Sri Aurobindo speaks of the next level of mind that we are capable of harnessing now. He referred to this as the supermind. While our current quality of rational mind provides a lens for Infinite Consciousness to experience itself as Matter in a very dualistic way, the supermind, or supramental mind allows Infinite Consciousness to experience itself within Matter while at the same time maintaining the awareness of primal unity.

As long as we are bound to the conditioning of the rational mind we will continue to experience ourselves as trapped within the collective matrix of duality. We are victims in a long chain of cause and effect, and too weak and powerless to extricate ourselves from it. We are insignificant pawns in a game where the laws of competition and survival of the fittest override the codes of co-operation, compassion, integrity and sovereignty.

However, as we open to the evolutionary possibilities of the supermind, this changes quickly and dramatically. We experience ourselves as infinite, powerful and creative beings lovingly creating a new earthly paradise based on the principles of harmony and beauty. I am not separate from my neighbor and therefore am no longer subject to the need to fear, betray or control anybody else. Just like each cell in the body works for the benefit of the whole, I find my unique gift and express it to support the whole, while being equally supported by the whole.

What this means on the human level is that we are not blinded anymore by the illusion of separateness and duality that veils us from the essential unity underlying all life and all things. We discover that our highest good is also the highest good of the planet; we find that Earth is home for all of life, and that we cannot misuse her resources to benefit just a few. We realize that our highest joy is to serve all life to the best of our ability, and that who we are is not simply an outer physical shell but the infinitely beautiful stream of all existence flowing through our open hearts, minds and bodies.

In other words, we discover that the Self in its role as creator becomes visible within bodies of creation. Sparks of divinity incarnate within the body as matter awakens to subtle realms of uncreated light. Subtle senses open within expanded perceptions of reality, breaking our hypnotic conditioning of a fixed reality 'out there', and flowing through multiple dimensions of time, space and existence.

As the supermind replaces our current dualistic mind, all evolution becomes a conscious dance of expression. A new species of humans serve their function as the nervous system of the much larger organism of Gaia, a biological collective that includes all life on Earth and even far beyond.

Within this biological collective, creative manifestation and joyful expression takes place in a constant act of new creation. The human species, whom Sri Aurobindo referred to as wide browed creators of eternity, take their place within the web of all creation as sun-eyed caretakers and architects of a new and marvelous dawn.

"A new world has already been born," says the Mother. *"It is not the old one transforming itself; it is a new world which is born. And we are right in the midst of this period of transition where the two are entangled – where the other still persists all powerful and entirely dominating the ordinary consciousness, but where the new one is quietly slipping in, imperceptible at first. And yet it is working, growing, until it becomes strong enough to assert itself visibly."*

As we perceive, so it is. The act of changing our perceptions of reality, changes reality itself, since reality is not independent from the one perceiving it. This is especially true when we are talking about the supramental quality of mind. Since the supramental force is birthed from the same space where the dance of creation happens, new realities can instantaneously be projected out through the lens of supramental vision.

This shift from rational mind to supermind is an evolutionary choice. Sri Aurobindo believed that it has already been made in wider fields of existence, and is rapidly being manifested within the field of matter. We can choose to be active pioneers in the embodiment of this new mind. It is not an external task, however, but needs to be accomplished within our own bodies, trusting and following the impulse striving to birth itself deep within the cellular mind.

"Tat twam asi," said the ancient Vedic seers. *"That I Am."* This is the realization sometimes referred to as enlightenment, the awareness that there is a single consciousness pervading all things, and this consciousness is not separate from what I am. The rational mind cannot be expected to understand this. The supramental understands nothing but this.

And from this arises the foundation for a New Earth.

CHAPTER 35
THE NOOSPHERE AND
THE NEW SPECIES

Sri Aurobindo refers to four stages of human evolution, culminating in the supramental human. This vision has been mirrored in the insights of many others within the past few decades.

Pierre Teilhard de Chardin was a paleontologist and Jesuit priest. In the middle of the last century, he proposed that humanity is evolving to an Omega Point, where a new consciousness — which has been percolating within an outer region of the Earth's atmosphere known as the noosphere — would be manifested. This would result in a new species of humanity capable of manifesting Christ Consciousness.

This concept of the noosphere links closely with Rupert Sheldrake's morphogenetic fields — intelligent fields of form that are capable of reproducing themselves. These form-generating fields govern the evolution of all species, as well as of Mother Earth herself.

There comes a time in the gradual evolution of a species when it is ready for a quantum jump. It is similar to when a nerve cell prepares to transmit information across an intervening synapse to a neighboring nerve cell.

Impulses accumulate at the synapse, until suddenly it pushes past a critical point at which the nerve cell fires, and information gets carried past the synapse to its destination. The hundredth monkey phenomenon, popularized by biologist Lyall Watson, is an example of what happens in a biological species when this critical mass is activated.

The information for the collective evolution of humanity is held within the morphogenetic field of the noosphere. It becomes available in incremental steps as we become ready for it. According to Teilhard, the noosphere is located in physical space within the ionosphere, which surrounds the biosphere of our planet. At this high altitude, it is extremely responsive to cosmic rays and gamma ray bursts in the early stages of their entry towards Earth.

An enigmatic phenomenon has been taking place over the last couple of decades that may well be related to new activity within the noosphere. This is the phenomenon known as crop circles which have been occurring all over the world, and especially in Wiltshire County in the south of England.

There has been much conjecture regarding who the circle-makers are. Leaving out obvious fakes (such as the Doug and Dave variety), the circles are attributed to sources ranging from extraterrestrial beings, to fairies, to a sort of collective Earth intelligence.

One theory for the origins of crop circles involves plasma tubes in the ionosphere. According to this theory, as shared by Geoff Stray in *Beyond 2012: Catastrophe or Ecstasy — A Complete Guide to End-of-Time Predictions*, the crop circles are created through an interaction between magnetized plasma bands within the ionosphere and information imprinted within the noosphere.

Our collective human memory as well as new evolutionary potential is imprinted as an akashic record within the noosphere. We could perceive it as a three-dimensional photographic plate for recording all the information that has ever existed in the past as well as genetic codes for our future evolution. Through the action of plasma tubes, this information can be out-pictured to create crop circles with the purpose of activating these codes in human and planetary consciousness.

Wiltshire is composed of mostly chalk, which can hold a lot of subterranean water. This creates a negative charge capable of attracting this form-creating plasma (similar to a lightning conductor). When the conditions are right, the noospheric plasma creates an intense magnetic field, which instantly manifests in the geometries of a crop circle.

In this regard, I would like to share a personal story. My interest in crop circles inspired a trip to the UK in July 2007. I wanted to experience them for myself. I had heard many stories of people coming together to visualize a specific theme which would show up as a crop circle soon afterwards.

I arrived in England on July 16th and immediately took a train down to visit some friends who lived not far from Stonehenge. We had a beautiful reunion, catching up on each other's lives. We then decided to experiment with asking the circle-makers to create a butterfly formation, which for me had become the theme of our collective awakening process. We remained in quiet meditation for a while.

I remained in the UK for a couple weeks afterwards. Busy with other things, I soon forgot about the seemingly failed experiment.

Two months later I was in the south of France offering a four-day seminar. One of the participants had a copy of a Nexus magazine, which included an 8-page color foldout of recent crop circles from the summer season. Imagine my surprise when I opened to a beautiful butterfly formation made exactly on July 16th in Hailey Wood, Oxfordshire!

Human bodies, like the chalk formations of Wiltshire County, contain significant amounts of water and are capable of attracting the same plasma imprints from the ionosphere. Any morphogenetic field of evolutionary possibility can therefore be received through the subtle energy fields of the human body.

Is this why there is such a spontaneous awakening happening for so many people around the world today? Could it be that our subtle energy bodies are being prepared to more easily receive these genetic codes from the evolving noosphere so that the morphogenetic field of a supramental consciousness can eventually descend into human form?

New possibilities are beginning to manifest all over the world through many different forms and many different traditions. They are also manifesting spontaneously outside of any tradition. Many people are starting to experience kundalini breakthroughs, spiritual openings, vivid dreams and visions, and cellular changes within their bodies, all of which signal an initial collective supramental descent.

The new generation is particularly open to these budding possibilities. Modern mythological heroes, such as Luke Skywalker, Harry Potter, Jake Sully, and the X-men, are opening doorways for them to trust in their own emerging gifts. Many teenagers and young children are exhibiting very unusual, and oftentimes challenging, perspectives and capabilities.

I often hear from so-called crystal kids, who are capable of doing 'miraculous' things with ease — affecting the weather, predicting the future, telepathy, healing people and animals, and communicating with inter-dimensional beings. For the most part, despite these gifts, they feel quite powerless and isolated in a world dominated by the limited perceptions of the rational mind. However, they represent the first phase of a new species.

"The heavens beyond are great and
Wonderful, but greater and more wonderful
Are the heavens within you.
It is these Edens that await the divine worker.
I become what I see in myself.
All that thought suggests to me, I can do;
All that thought reveals in me, I can become.
This should be man's unshakable faith in himself
Because God dwells in him."

The Mother

CHAPTER 36
EMERGENCE OF THE BUTTERFLY

I have a friend in Colorado whom I will call Joe. At one point in his life, he underwent an intense exploration of Eckankar, a system of soul travel. We often met to share stories and perspectives of life. One day he shared with me some unusual events he had experienced during a period of profound spiritual opening.

Walking down the street, thinking about a girlfriend in New York, he found himself entering a profoundly altered state of consciousness. Suddenly, he discovered he was on the streets of New York. It took him some time to acclimate to this new reality, but just as he was getting used to the fact that he had somehow transported himself in his physical body, he realized he did not have her address, nor money or credit cards on him.

Then, just as suddenly, he was back in Colorado. For the next two weeks he remained in a state of consciousness where he could think of going someplace, and start walking. At a certain point a dimensional doorway would open up and he would be transported there.

One day he decided to see if he could take somebody else along. He invited a friend to do an experiment with him. His friend agreed, but in the moment of engaging the 'transporter beam', he became frightened and pulled back. In fact, he was so shaken by the experience that he turned on Joe and had him committed to a psychiatric ward. In order to get him back to 'normal' they placed him on anti-psychotic drugs, which unfortunately destroyed the delicate neurological pathways that made such transportation possible.

Another friend of mine, Julian, was stopped by the highway patrol one day. They found a small amount of marijuana on him and carted him off to jail where he was to serve a three-month sentence. He had been practicing Drunvalo Melchizedek's 'merkaba meditation' for activating the light body.

As he continued to do the merkaba practice in his jail cell, he suddenly found himself out in the woods near his home. Petrified that the authorities would come after him as an escaped prisoner, Julian hid out. Just before his sentence was to expire, however, he found himself back in jail. As shocked as he was, the guards were even more so, and interrogated him about where he had been. Further investigation, though, would have taken them into unfamiliar and uncomfortable territory. They chose to let him go.

I hear many such stories as I travel, including stories of incredible healings. My close friend Patricia once attended a forty-day healing camp near Mt. Shasta, California. She was coming down the mountain with some friends when their car went out of control around a sharp bend. It collided with a tree and everyone was thrown out.

As Patricia hovered in and out of her body, her spine broken in several different places, she saw a policeman approaching her. *"Do you want to live or die?"* asked the policeman, whom she later identified as possibly St. Germaine. She said it didn't really matter. She was happy to go, but if there was something she still needed to do on Earth, she would stay. *"So be it,"* he said, and disappeared.

Patricia reports that a half-hour later the real policemen showed up. They airlifted her to Sacramento where the doctors examined her injuries. They told her she would never walk again. As the nurses were about to lift her from the gurney to her hospital bed, Patricia warned that she would probably scream in the intensity of her pain. Just then, very dramatically, the door flew open. An African-American man with a wild afro hairstyle in a nurses jacket demanded that everyone step back.

Introducing himself to the nurses, he explained that he was new on the floor that shift. He told Patricia not to worry; he was working on a new, pain-free technique for lifting patients. He was even writing a book on it. Gently, he picked her up like a baby and put her on the bed. Indeed, she felt no pain.

He visited her daily. He never took her blood pressure, her temperature or did any of the routine tasks of a nurse. He only talked to her about God and about the meaning of life on this planet — to love. In fact, he said, the very nature of our existence is love.

Although she was expected to stay in the hospital for up to six months, she was out in two weeks. When she returned to the hospital to thank this nurse whom she called her angel, she was informed that he had left the very same day that she left. No one could remember his name nor did they know where he went.

A few years later when Sai Baba came into her life, she recognized his picture and realized that it was he who had shown up as the nurse in the hospital. His presence had created an enormous vortex of healing. Her body, incredibly, had taken on so much healing light that it became tanned! When I met her a year later, she was walking, running, and skiing without any trace of injury or pain.

I believe that these kinds of phenomena are much more widespread than we realize. Many people have had these kinds of experiences but don't know how to interpret them. Therefore, they either repress them from conscious memory or refuse to share them with others out of fear of being considered crazy.

The morphogenetic field of possibility is growing. Our experience of time is changing and our ability to experience higher states of consciousness is increasing. As we become increasingly receptive to higher frequencies of energy, however, it often feels as though we are losing our minds and our memories. This also has to do with the ongoing collapse of the geomagnetic fields.

As I travel with my seminars, I often ask how many people feel this way. Usually, every hand goes up. This sense of losing our minds seems to be an integral element in the process of transformation. We are literally losing our minds in order to arrive at our senses. We are beginning to move out of linear consciousness into multidimensional consciousness. However, the no-man's land in between can be quite disorienting.

A beautiful example of this is the journey of a caterpillar as it becomes a butterfly. The caterpillar is happily munching away on leaves when suddenly new cells, known as imaginal cells, start to form. The caterpillar's immune system tries to protect itself from the invading cells, but there are soon too many.

Overwhelmed, the caterpillar creates and enters a cocoon. Its DNA transforms as these imaginal cells take over. The cells reorganize the entire body until eventually the butterfly template is created. One day the butterfly breaks out of its cocoon on wings it has never before known.

Our bodies are going through a similar process in response to the galactic energies beginning to bombard our planet. The timing is different for each person, but, in some way, we are all starting to experience this transition into a cocoon stage. New DNA codes are being activated, and each of us is being called to transition through the birth tunnel of a new world.

Many spiritual channels talk about a conversion to 12 strands of DNA. Although I have yet to see any scientific evidence for this, I do feel it is true that previously dormant genetic codons are being turned on.

Scientists tell us that only 3-5 percent of our DNA is actually useful. The rest is considered junk DNA. What if this isn't just a waste pile of genetic material, but rather the raw material for creating the divine human? Some biologists are considering that this so-called junk is actually a set of dormant programs. Sri Aurobindo referred to us as a transitional species. Could it be that we are individually and collectively preparing to enter through the cocoon and emerge as a human butterfly?

A caterpillar's journey through the cocoon stage is not easy. It involves a complete dissolution of its previous consciousness and even its physical body. Likewise, we can expect to experience compromised immune systems and various levels of physical and emotional stress as our own DNA transforms.

Although this is happening spontaneously to people everywhere, it may be especially true for those actively pursuing a spiritual path and practicing to awaken their kundalini — the subtle currents of life energy that lie dormant in most of us, and that once activated begin to unify soul consciousness with physical consciousness.

Possible symptoms of this process of awakening include headaches, joint pains, heart palpitations, dizziness, frequent fevers, hot and cold flashes, immune system breakdowns, states of extreme tiredness or extreme restlessness, psychic openings, and vivid dreams. Sometimes these symptoms are accompanied by feelings of cosmic union; other times by feelings of intense agony and separation.

All of this is an essential stripping down of an identity based on the rational mind, or personal ego, in preparation for our journey into the divine human, and perhaps many generations later, the supramental human.

This physical and emotional clearing is sometimes known as the dark night of the soul, and serves to set us free from outer dependencies. It puts us in touch with an inner process where the mind begins to let go of limited perceptions, false conditioning, and illusionary certainties. As we allow our souls to take over, deep transformation begins.

For those who may be familiar with astrology, Pluto entered into Capricorn in February 2008, and will remain there until 2024. This implies a breakdown and transformation of all forms and structures, including physical bodies, planetary ecosystems, political systems, and the global economy. It will be interesting to observe how we experience this transit in our journey onwards.

I referred earlier to the theory that crop circles are created through the activity of magnetized plasma as it is imprinted with intelligent information from the noosphere. What happens to the wheat growing inside these crop circles? Geoff Stray, in an appendix to *Beyond 2012*, refers to the long-term effects of this magnetized plasma on crops within the vicinity of a crop circle.

In an immature crop, he points out that the plant develops normally, but never develops any seeds. In a young plant in which the seeds are already forming, it will develop smaller seeds with reduced germination. But in mature plants with fully formed seeds, the magnetized plasma generates a massive increase in growth rate and in vigor. If we apply this to the dimensional transformation that we are about to undergo as a species, it may be an example of how the increase of incoming cosmic rays can affect us at different stages of spiritual preparation. Perhaps it points to a mechanism whereby different dimensional realities could be created.

Those who are spiritually immature, locked into a worldview based on ego-dominance, would move into a corresponding dimension of conflict and catastrophic outcome. Those between would experience something, well, between. But the greatest increase would come for those whose hearts, minds, and bodies could effectively use the energy to transition into an ascended level of consciousness.

This next species of humanity has been called by different names by different people. Teilhard de Chardin referred to the christed human, while Sri Aurobindo speaks of the divine human. Barbara Marx Hubbard calls this new species homo universalis, while Peruvian shamans tell us we are birthing the homo luminous. Others refer to galactic humans, quantum humans, and the sun-eyed children of a marvelous dawn.

We are probably closer to this new species than we realize. The birth tunnel is dark, painful, and chaotic. Regardless, there is a power pushing us forward that is beyond our ability to understand.

There are at least two stages of human evolution still to come. Beyond the divine human is the supramental human. As Georges van Vrekhem quotes the Mother, the supramental human is *"a true being, perfect in proportion, very, very beautiful and strong, light, luminous or else transparent. It will have a supple and malleable body endowed with extraordinary capacities and able to do everything; a body without age... a transformed body such as none has ever imagined."*[1]

A supramental body is composed of substance that is different from ordinary matter, emphasized the Mother. Supramental substance is charged with supramental consciousness. It is not susceptible to aging, disease, or death. While carrying a greater density than ordinary matter, it is subtle, supple, and refined enough to hold the enormous power of creation. It can travel from place to place instantly and allows one to be present in several places at the same time. It concretely expresses divine attributes such as omniscience, omnipotence and omnipresence.

A supramental body operates according to entirely different laws of the universe, much more akin to what science has attributed to the world of quantum physics. Here, finally, is the realm where all experience of duality and separation becomes erased within the field of matter, where creator and creation become one. This, according to Sri Aurobindo and the Mother, is what human evolution is heading towards!

CHAPTER 37
GROUNDING THE VISION

Our journey from caterpillar to butterfly takes us through the chrysalis, an in-between stage where we are neither here nor there. The big picture of cosmic evolution is useful as we travel through this birth tunnel. We are not yet there. We must take clear and immediate steps in the here and now to prepare for this. How do we ground this cosmic vision in the context of our present-day world?

When we can hold a clear grounding in our current reality, just as it is, and also a profound vision of highest possibility, a strong dynamic tension stimulates the creative process. Most people tend towards one or the other of these two poles of the creative process and thus minimize their effectiveness. In order to become agents of change, we must have both a solid footing in current realities and a clear vision of future possibilities.

A vision of possibility without a strong grounding in current reality leads to passive ungrounded dreaming. Conversely, a strong grounding in current reality without a clear vision results in stagnancy. In such a state there is nothing to hope for and nothing to move toward. One is left endlessly spinning in the realm of the problem. Can we instead look at the big picture of planetary evolution and use it to fuel new possibilities in our current response to global crisis?

If we know the nature of the energies moving towards us, can we assemble our leading visionaries, scientists, and most dedicated policy makers to harness this energy, rather than being helplessly inundated by it? If we understand the mechanics of the transformation at hand, could we not prepare ourselves as individuals more effectively to surf the wave?

With the exception of a small group of influential people still locked into control dramas, I believe that most of humanity is rapidly tiring of a world controlled by greed, oppression, and war. More people are responding to the call of their souls. More are starting to discover the big picture. Each of us holds a thread in the grand loom of times to come, one note in a beautiful song of awakening. How do we weave these together? We are preparing for a quantum leap in our evolution that will give rise to a unified world soul — a global brain within the body of Gaia. How can we facilitate this emergence?

Many years ago, when I still lived in Mt. Shasta, my friend Barry Martin and I conceived of a chrysalis network for catalyzing the emergence of this new world.[1] Over the years we have continued to collaborate on manifesting this vision. In an increasingly interconnected world, it seems obvious that we cannot remain isolated and evolve further.

Quantum physicists need to speak their deep understandings of natural law within the structures of global politics; industrialists and politicians need to become fully aware of the impact of the free energy revolution on shaping a new global economy; inventors need to appreciate the impact of such technologies on healing our environment; environmentalists need to work more interactively with policy makers; and healers need to understand the implications of new medical research.

It is important that a wise new spirituality, relevant to the challenges and opportunities of these times, emerges among us to usher us beyond the confines of religious dogma into the dawning Age of Light.

Likewise, new paradigms in childbirth, parenting, business, education, architecture, farming, community building, healing, psychology, science, technology, and politics need to be explored more fully. Inspired visionaries need to be funded by philanthropists who have the good of the planet at heart, and inspired ideas need to be disseminated through music, art, gatherings, books, and the media networks in a way that can reach the masses. Millions of cultural creatives around the world, each with their own unique gifts and skills, are waking up. They are coming together as one heart and mind under the guidance of a single, emerging planetary soul that seeks to inspire our collective potential.

We have discussed the possibility of a huge dimensional shift to come. It is by no means certain what this will look like or how we can best prepare for it. An essential role of this global think tank — which includes geologists, astrophysicists, politicians, futurists, spiritual elders, shamans, and historians — would be to guide our passage through these times of transition with grace, ease, and understanding.

Earlier in this book, I attempted to make a connection between the spiritual aspects of LaViolette's galactic superwave and Sri Aurobindo's supramental light. As conscious beings, our own choices and perspectives make all the difference. I believe that this process involves feedback loops between collective humanity and a galactic intelligence, which shape our physical encounter with this galactic phenomenon. Our collective density level will determine how we experience the chain of effects following the coming galactic superwave.

It may be too much to hope that the whole of humanity will be ready for this superwave in time to avoid at least some degree of catastrophe. However, I derive hope from the findings of Dr. David Hawkins, a psychiatrist who devised a method capable of measuring the frequencies of human consciousness through applied kinesiology. Hawkins' scale of consciousness spans 0 to 1000, with a turning point around 200 where we move from self-destructive to life-enhancing traits and tendencies.

At the bottom of the scale is guilt and shame, progressing up through apathy, grief, and fear, to anger, pride, and finally, at level 200, to courage. Above this line, beginning with courage, we ascend to neutrality, acceptance, reason, then love, joy, and peace. Finally, from level 700 onwards, we explore various stages of enlightenment. The journey up this scale is what constitutes our spiritual evolution.

Even though the average human consciousness on the planet today is still below 200, the collective human consciousness is above this line. This is because each person vibrating to higher consciousness on this scale counterbalances many others on lower levels of consciousness. The scale advances logarithmically. Thus, one person vibrating at love, 500 on this scale, counterbalances 750,000 individuals below the line; while one enlightened person at 700 can counterbalance 70 million below the line!

As we connect with the galactic consciousness within our own being — whether individually or in think tanks and small groups — we begin to activate these feedback loops. I believe that we are fast approaching critical mass on this consciousness scale. Once we do, we can use this enormous wave of evolutionary energy to collectively birth ourselves in a quantum leap of consciousness. Are we ready?

PART IV

TRANSITION

Our lives, our past and our future are tied to the sun, the moon and the stars... we, who embody the eyes, the ears, the thoughts and feelings of the cosmos, have begun to wonder about our origins, star stuff contemplating the stars, organized collections of ten billion billion billion atoms contemplating the evolution of nature... Our loyalties are to the species and to the planet. Our obligation to survive and to flourish is owed not just to ourselves but also to that cosmos ancient and vast from which we spring.
We are one species.
We are star stuff harvesting star light.

Carl Sagan

CHAPTER 38
AN EVOLUTIONARY LEAP

In the past few chapters we have been laying the theoretical groundwork for a profound planetary shift. Amidst ecological ruin, financial collapse, political breakdowns, personal traumas, and social upheavals around the world, there is a clear powerful evolutionary process at work.

We have looked at this evolutionary process in terms of long-term cosmic cycles of approximately 12,000 years. We have seen how galactic superwaves, long-term solar cycles, ice ages, and magnetic reversals all seem linked together in this evolutionary process.

We have seen how a gradual period of evolution is usually followed by sudden and dramatic quantum shifts as we move from one cycle to the next and one world age to another, roughly corresponding with Yukteswar's model of the four yugas. We have seen how mystical consciousness can play a part in predicting, mapping, and facilitating these changes. We have examined Sri Aurobindo's vision of the supramental descent as a key element in a co-creational process of birthing the New Earth.

As we look back over the past few years it's very clear that things are accelerating. Fluctuations in the magnetic field are getting stronger. This is evidenced by recent phenomena such as thousands of migratory birds dropping dead from the skies, entire colonies of bees disappearing, and the periodic beaching of vast numbers of whales and dolphins.

The fluctuation and drifting of magnetic poles is usually a prelude to magnetic reversal. At the rate of approximately 40 miles a year, the magnetic north pole is drifting from its current location in North America towards Siberia. Airport runways around the world have had to be continually recalibrated to reflect this drift.

As we move closer to a magnetic reversal, it is likely that we will simultaneously experience a reset in human and planetary consciousness. As Felix has described, there is a link between magnetic reversals, ice ages, and evolutionary leaps. The stage is already set for massive changes ahead — the magnetic reversal is already starting to happen, a mini ice age has already begun, and an evolutionary shift will inevitably follow.

Don Alejandro emphasizes that although we do know it will be soon, we cannot predict the timing of Year Zero as an exact date. It is much like predicting the birth of a child.

Birth is a beginning, not the end of a cycle. Once a child is born it takes a lifetime to fully mature. Year Zero represents not so much the end of the world as the beginning of a new cycle. How can the new consciousness emerging from the Shift shape the new world to come?

It is truly a miracle that life on our beautiful planet exists at all. If our planet were a little further away from the Sun, it would be too cold to sustain life. If it were just a little closer we would burn to a crisp. If the force of gravity were a little stronger, then galaxies and planets would not have differentiated from the original cosmic soup. A little weaker and we would be drifting out into the coldness of space forever.

There are hundreds of such narrow parameters that shape our existence within the fabric of life. As James Lovelock, Peter Russell (*The Global Brain*), Rupert Sheldrake (*A New Science of Life*), and other scientists have affirmed, the very fact of our existence here on Earth is truly a miracle. The probability that we as humans could have evolved simply by chance to the point where we could be asking ourselves this question is in the order of 1050, a mind-boggling 10 followed by fifty zeros!

So how did all this happen? Is there a God, some Great Cosmic Architect, who created all of this in the beginning of time, including all the laws of the Universe, and set it to run like clockwork? Did it all happen randomly or through some yet unknown principle of evolution?

There is a new theory of everything proposed by physicist Robert Lanza known as biocentrism. In *Biocentrism*, Lanza and co-author, Bob Berman, assert that the creation of the universe, of space, time, and matter, did not happen by chance or by a single act of creation. We, the observer, create time and space and the laws of nature. We create not only what is yet to come, but also our perceptions of what has past.

In esoteric language we could rephrase this to say that we ourselves are the creator gods through which evolution happens. Our higher selves are one with creator consciousness, and we are constantly creating new possibilities through our unified thoughts and intentions.

LaViolette, in his theory of continuous evolution, equates new creative possibilities with the active phase of the galactic heartbeat. In this active phase massive amounts of cosmic and gamma rays are ejected from the galactic center in periodic bursts. The rays stream out through space as galactic superwaves. Could it be that the next galactic superwave is already beginning to move through our Solar System, initiating the physical events leading to an intense acceleration in our evolutionary potential? Is this what Sri Aurobindo's supramental descent is all about?

And if we are birthing a new world, how does this effect the daily functioning of our lives here on Earth?

In these next chapters I would like to explore how the evolutionary force moving through our planetary consciousness might be influencing our sciences and technologies, our religions and psychologies, our social relationships, and our economic and political systems.

Maybe there are no fixed outcomes — no single set of probabilities for the future. We can choose to create whatever timelines we wish based on the infinite power of our creator consciousness. Our beliefs and perceptions shape our reality. As we envision, so we create. The question is not so much what kind of world we will passively inherit, but what kind of world do we actively choose to create.

CHAPTER 39
GAIAN TECHNOLOGIES

There is a direct relationship between what fuels us on the inside and what fuels us on the outside. The petroleum industry provides for most of our energy needs today. Harnessing ancient sunlight trapped underground from a million years ago, petroleum is non-renewable, polluting, and expensive. It concentrates power in the hands of the few and leads to competition, corruption, and global strife. As a source of energy, it is entropic—it creates increasing chaos as it breaks down.

There are other forms of energy that are clean, renewable, and inexpensive. What is more, they are syntropic, working in harmony with universal principles, and therefore keeping the cycle of creation balanced. There are three categories of syntropic energies I would like to discuss here.

The first category consists of various forms of increasingly popular alternative power sources such as sun and wind-generated electricity. Progressively efficient forms of solar cells are being developed every passing year. More and more automobiles are powered by clean fuels such as CNG (compressed natural gas). Electric cars, hybrid cars, and even water-powered hydrogen burning cars are becoming available. Wind, tidal and geothermal energies are being harnessed. When most people talk about alternative energy sources, this is the category they are referring to.

Then there is the category of free energy devices. Barely a few decades ago the brilliant Yugoslavian-born scientist Nicola Tesla offered the world a technology based on tapping the energies of the ionosphere. This meant that each person, using his or her own conversion unit, could power all their energy needs (at least as far as stationary power went) without cables, poles, or monthly bills.

Tesla's work was largely suppressed by power companies, oil companies, and various other controlling interests. Then came zero point technologies, hydrogen fuel cells and cold fusion, bloom boxes, plasma converters, and global scaling, all based in the renewed understanding of mysterious new energies that modern science was beginning to discover.[1]

I once knew an inventor in Florida who had developed a gravity machine capable of converting gravitational energy into kinetic energy. A very portable unit, he used it in his home to run his appliances and in his car when he wanted to go for a drive. While working on his prototype, he discovered that the prostate cancer he had been suffering from suddenly disappeared. He wondered whether this may somehow be connected to his gravity machine and invited some of his acquaintances over who had been suffering from myriad other ailments. He found, incredibly, that these too soon disappeared.

Why were these technologies never made public?

There have been innumerable inventors over recent decades whose discoveries have been systematically suppressed, whose laboratories have been destroyed, and sometimes, whose lives have been terminated, in order that such technologies not compete with the power and oil monopolies. Greed, as well as the lust for power and control certainly have a role to play in the suppression of free energy technologies.

There are other reasons as well. Anything that is manifested out into the world has its source and reflection in human consciousness. As long as collective humanity is held in separation and functioning below the 200 mark on Hawkins' scale of consciousness, we cannot help but out-picture our own inner chaos and imbalances in the natural world — gobbling up natural resources, polluting the Earth, and creating entropy wherever we go.

Where do we go from here? If we are talking about a time of Shift — a time of embodying spiritual consciousness in bodies of matter — can we expect something altogether new to emerge in the field of technology as well?

As mentioned, Einstein said that we cannot hope to solve a problem from the same level of consciousness that created the problem. But what happens when we begin to awaken the global brain where we expand human consciousness to merge with planetary consciousness?

I believe that once we begin to awaken to our divine creator potential, the free energy revolution will quickly replace our current primitive sources of power and energy. Not only that, we will begin to connect with a third category of technologies — what I call Gaia technologies — which will allow us to begin creating heaven on Earth.

Lovelock's Gaia hypothesis is based on his understanding that the Earth functions as one single, interdependent, living organism. This is true on more levels than we can physically observe. I have come to experience that the web of life we know as Gaia extends into non-physical levels as well.

Most indigenous people around the world still honor the invisible forces of the natural world that we generally refer to as nature spirits, devas, fairies, elves, archangels, dragons, or Elohim. These are forces that do not necessarily have a place in conventional science but which readily make their presence known to those who are willing to seek them out.

As we develop a relationship with these forces, we can embark on a journey of co-creation that could rapidly reverse the ecological destruction of our planet and reveal new possibilities for fulfilling our evolutionary potential.

It seems to me that humans act as cosmic conductors in the planetary web of life. When we cut ourselves off from divine light, we become a cancer on the face of the Earth. However, as we connect with this divine potential we can help Gaia evolve to her own highest potential as well. Our nervous systems are capable of accessing higher vibrations of light for the benefit of all creation.

What happens when we work in harmony with the invisible forces of nature? Communication with garden spirits through communities such as Findhorn and Perelandra are increasingly well known today. More and more people are also beginning to experience the enormous potential for Earth healing in their interactions with vast elemental presences that we might refer to as archangels, Elohim, or dragons.

The movie *Avatar* is a beautiful portrayal of a race of people who, living in profound harmony with nature, have become dragon riders. They have learned to interact with elemental energies and to merge their essence with the web of life. This is so different from our current culture where we are trained to fight and kill our dragons, which are the symbol of our own feminine beauty, and the power of the Earth.

We may find that once we experience this connection with the web of life, all the outer technologies we have come to rely on are pale reflections of corresponding inner technologies that are clean, syntropic, virtually unlimited Gaian technologies.

Our current exploitative technologies are entropic in that they increasingly generate chaos. As we discover our inner unity, our technologies become syntropic. As with the inventor of the gravity device, we begin to discover and apply natural laws that may seem miraculous to us today.

The Internet, for instance, is simply the reflection of a vast multidimensional network of information we could refer to as the innernet. Once we begin to connect with higher levels of mind, we enable a connection with these holographic records hidden within all creation. This includes memories of the past and of the future. These higher dimensions exist outside of time in a realm where past, present, and future merge.

When we connect with higher dimensions, we discover that each of our outer senses has a counterpart in subtler dimensions. This is the basis for phenomena such as precognition, telepathy, psychokinesis, bilocation, shapeshifting, and ascension. Similarly, all outer technologies are reflected in corresponding subtle technologies. As our bodies begin to vibrate to higher frequencies of light, we will begin to discover how these Gaian technologies manifest in the world around us.

Drunvalo Melchizedek tells us that high technology is not the sign of an advanced civilization. It is the sign of a civilization that is about to be advanced. What good is a technology to a people who discover that the human body and human consciousness is capable of doing everything that that technology is now doing, and far, far more?

Perhaps we will discover that we can directly absorb prana — the living field of light that permeates all things — and no longer need food to nourish our bodies. We may learn to control our internal temperature by simply adjusting the kundalini fires running through our bodies. We may discover that we can teleport our bodies anywhere we choose bypassing the cumbersome means of transportation we currently use.

Perhaps we will find that as we begin to understand and activate the vast intelligence within our bodies, we will make unprecedented breakthroughs in the fields of longevity and health. We may discover that we can work with the archangels, dragons, and elemental spirits of nature to quickly and easily reverse the ecological destruction and nuclear devasatation we have created in our long cycles of ignorance, greed, and separation.

As conscious creators we are no longer subject to the laws of entropy and death. Instead, we can truly embark on a voyage of discovery to explore strange new worlds, to seek out new life and new civilizations, to boldly go where no man has gone before!

CHAPTER 40
SOUL PSYCHOLOGY

There is a longing for life within all matter and a longing for the full expression of itself within all life. This longing, call it what we will, is the impulse that drives our entire journey of involution and evolution. Spirit descending into matter is involution. Matter rising towards Spirit is evolution.

When we are identified with our human personality, our longing is towards the realm of Spirit, the transcendence of our material limitations, the fulfillment of our deep yearnings for divine expression. When we are identified with our divine soul, our longing is to explore the worlds of matter, expressing our multidimensional presence in the worlds of form.

What we are is neither body nor spirit, but a merger of both, expressing as human personality as well as divine soul. From the perspective of our limited, human self, locked in a world of duality and dense physical body, there is a force that drives us towards the full embodiment of our vast soul presence — our journey of evolution. And from the perspective of atman, our spirit self, our challenge is to be fully embodied in the world of matter — our journey of involution.

We are at a stage of existence where soul and personality are experienced as separate. Our task is to merge them into union.

The task of soul psychology is to give us an understanding of what this means and the tools necessary to achieve this. The task of spirituality is to experience this union and to achieve mastery on all levels of consciousness.

Let's begin with the experience of separation. The kahunas, or priests, of Hawaii refer to three aspects of the self — the high self is the super-conscious, the middle self is the conscious personality, and the low self is the subconscious. Their spiritual practices are designed to unify these three aspects of self.

The kahunas assert that the middle self cannot communicate directly with the high self. In other words, our conscious efforts and intentions, no matter how sincere or strenuous, cannot put us in touch with the high self if the low self doesn't cooperate.

The low self, on the other hand, is in constant touch with the high self, although the quality of this connection may vary. Our task is to make friends with the low self, sometimes referred to as the inner child.

Western psychology is generally focused on the wounded aspects of this inner child. Sigmund Freud's psychoanalytical system was an early effort to understand the neurotic tendencies of our hidden subconscious and to delve into the shadow realms of the ego. Carl Jung took this further. Jung recognized that the shadows were simply the reflection of the light of our inner soul. He explored the larger archetypal realm underlying our personal subconscious — which is both dark and light.

The contributions of transpersonal psychologists Roberto Assagioli, Ken Wilber, Stanislav Grof, Arnold Mindell, among others, have refined our understanding of the superconscious aspects of the soul and the subconscious aspects of human personality in an attempt to bridge the two. Various techniques and awareness practices, including dreamwork, bodywork, breathwork, shamanic journeying, and inner dialogue facilitate and deepen the connections among the high self, middle self, and low self.

Although all these methods have contributed significantly to the understanding of our selves in human embodiment, I would like to take some time now to explore even subtler realms of body and mind as we prepare for the next stage of our evolution/involution into the divine human.

Let's go back to our involutionary journey 13.6 billion years ago.[1] We existed then as undifferentiated consciousness, preparing to create a universe where we could experience and express ourselves in the multiplicity of form. The Big Bang happened—the fabric of time and space was created, resulting in an explosion of energy and matter.

This Big Bang was our moment of involution, our initial entry into the world of matter. It has taken 13.6 billion years for creation to reach its current stage of evolution. We are just recognizing our true face within the dance of matter and preparing to embody the consciousness of creator in the realms of creation.

How does this journey actually happen? From a spiritual perspective, it is not random chance that drives us onwards. Rather, it is an evolutionary impulse within the heart of creation. This divine impulse within matter has led to the creation of universes, galaxies, suns, planets, oceans, and life. This same impulse has shaped living organisms into increasingly greater complexity. This impulse has, in the fullness of time, formed bodied species with nervous systems capable of becoming vehicles for creator consciousness.

Soul increasingly embodies as life evolves. Our evolution is by no means complete. While it may be that there are species on our planet more evolved than we are, humans are yet a transitional species. If we are to continue into the next stage of our evolution as divine humans, it may help us to get an idea of what lies ahead.

It is obvious to us that we have physical bodies. Yet we also have various subtle bodies invisible to the physical senses that begin to develop as we continue to evolve further.

The ancient Egyptians had a clear understanding of these subtle bodies. They referred to the physical body as Khat and were trained to work with its etheric counterpart known as the Ka. The ka included an extensive network of energy meridians that nourished the khat whilst extending beyond it. This network of light allowed initiates to perceive higher-dimensional worlds.

Beyond this was the *akh* — a luminous body of light that we might also refer to as the astral body. With training, this body could function independent of the physical body, traveling instantly through time and space. With practice, the initiate could consciously attune to this luminous body and use it as needed.

The soul was known as the *ba,* and there was a body of light directly associated with this known as the Sahu, which we might also refer to as the causal body. Initiates who learned to work with the sahu could receive the light of the soul directly into the physical body, thereby reversing the aging process and experiencing immortality.

These various layers of the subtle body are also linked with various levels of the mind. Sri Aurobindo refers to five levels of the mind beyond the thinking mind.

The thinking mind is the rational mind — used for gathering and analyzing concrete information and mediating the physical senses — this is the level of mind we are most familiar with. However, many levels of mind beyond this serve as gateways to multidimensional consciousness.

The higher mind is where we first become conscious of our spiritual self. We begin to perceive and interact in the world through this expanded consciousness. It is the realm of inspiration. If we compare the thinking mind to a twilight zone, the noonday Sun becomes clearly visible to us in the higher mind. It corresponds with the ka body of the ancient Egyptians.

Next comes the illumined mind. The light of Spirit breaks through into physical consciousness, illuminating the mind, heart and body with its luminous power. Like a bolt of lightning streaking through the skies, we see clearly in all dimensions whatever knowledge we seek. The luminosity of this plane of consciousness can be likened to the luminosity of the akh.

Beyond this is the intuitive mind, where we simply merge our consciousness with that which we seek. This is the realm of direct intuitive transference. There is no more duality; no more involvement of the senses while perceiving truth. We simply become the truth we seek. This plane of consciousness is associated with the sahu.

The overmind is an archetypal realm. It exists beyond time and space, as does the ba, in the creative realm of the gods where all is possible. From the perspective of the soul, it is the first stage of our descent into the created universes.

Beyond the overmind is the supermind — which perceives all truth in absolute harmony and unity. This is the gateway to the undifferentiated consciousness of the Great Beyond. It is the level of the mind that Sri Aurobindo and the Mother were striving to anchor into the realms of matter.

This is the realm of a next stage of psychology, beyond psychoanalysis, behavioral psychology, depth psychology, humanistic psychology and transpersonal psychology, which is referred to as Integral Psychology. It is a realm of psychology that requires psychological as well as psychic tools, and moves towards emotional as well as spiritual maturity, integrating the different stages of our evolution and the different levels of the mind.

Evolution is simply about being aware of who we already are. Each level of Mind is conscious of that level of Self. As supramental consciousness descends into the slumbering heart of matter, it opens the doorway for a mighty shift in consciousness where creator can finally awaken within the heart of creation!

The Supreme Height
Touches the most material matter
All the splendors one can experience
By going up, by getting out,
By leaving are nothing!
They're nothing; they don't have
That concrete reality;
They seem vague compared to HERE
That is truly why the world has been created
It's in terrestrial matter, on Earth
That the supreme becomes revealed.

CHAPTER 41
AWAKENED SPIRITUALITY

If the task of soul psychology is to understand the multi-dimensional aspects of our being, the task of true spirituality is to embody this.

Many enlightened masters have been springing up around the world in recent decades to share their teachings and perspectives with a growing multitude of seekers. The trend seems to be away from religious dogma towards a direct experience of the divine. In response to a deep stirring of soul within the unconscious depths of matter, humanity is beginning to awaken to the realization of who we essentially are.

In my own life, this search for deeper spirituality has taken me through an exciting journey of exploration. There have been many gurus, teachers, guides, and shamans who have provided revelation and insight at various times in my life.

I travelled to ashrams, monasteries, sweat lodges, shrines, and sacred sites around the world. I read every spiritual classic I could find. I learned to communicate with higher-dimensional intelligences. I trained to become a transpersonal psycho-therapist. I experienced intense emotional clearings. I spent time in the jungles with sacred plant medicines. I swam with whales and dolphins. I muddled through relationships.

All this has taught me much. I learned to not become too attached to any one version of truth. I learned that life is a precious gift to honor and cherish. I learned that the Earth is mother to all things, and that all creatures have an equal right to exist. I learned to look for hope in the midst of chaos and despair. I learned that love is all that really matters. I learned from mistakes and extremes. I learned to look at the big picture. I learned to acknowledge and embrace my shadows, for they are simply a gateway to light. I learned that we are on a journey of endless possibilities. And I learned that I don't have to walk this journey alone.

There have been teachers in my life that I will love and cherish forever. Jesus, Babaji, and St. Germaine have been among my inner teachers and guides, along with Sri Aurobindo and the Mother. They have held me and inspired me and taught me much about the journey into oneness.

Each of us has a journey to make, and each of us will be guided by our own unique destiny. Don Juan speaks of following the path with a heart. Joseph Campbell spoke about following your bliss. Each of us is guided from within to find the path that has most resonance for us at any given stage in our lives.

When we listen to the guidance of our hearts, we may find that we are sometimes called to step out of the box and to break out of the routines of daily life. Whenever we are ready to move on there is always a teacher or guide who shows up to inspire and support us along the way. It is a journey of trust. The universe can hold us and carry us always in safety and grace towards the fulfillment of our highest destiny. Are we willing to faithfully jump?

Ultimately, the teacher we must learn to trust most is our own higher self. As we prepare for this time of the Great Shift, the veils between the worlds are thinning. Many are finding it easier than ever before to access the higher levels of the mind — the source of this inner guidance. Once we learn to access our higher self, the next step is to integrate it.

In times past, the goal of the spiritual path was trans-cendence. The physical body and the material worlds were considered an illusion or a burden. The goal of enlightenment was to get rid of our egos, detach from the material worlds, and enter into cosmic consciousness.

This is changing now. With the descent of supramental light comes a new possibility of transforming the physical body and the world of matter. This means fully entering into the material worlds, descending into the subconscious depths of our ego-mind, embracing the hidden roots of duality and separation, and thereby transforming our deep-seated trauma, con-ditioning, fear and pain.

"In the spiritual tradition," says Sri Aurobindo, *"the body has been regarded as an obstacle, incapable of spiritualization or transmutation… But if a total transformation of being is our aim, then a transformation of the body must be an indispensable part of it. Without that, no full divine life on Earth is possible."*

We are at a stage in evolution where spiritual consciousness can be birthed within the realms of matter. Once this happens, the evolutionary process is directed not from a blind instinctive impulse within the density of matter, but in conscious participation with the realms of Spirit. A divine intelligence begins to awaken within the cells of our bodies to accelerate our journey onwards.

This is not simply a spiritual evolution but also a biological evolution. Most of our DNA material is relatively unorganized. Remember, only about 5% of our genetic material is turned on. The rest of our DNA's potential exists in a slumbering, supposedly junk, state. Microbiologists like Fritz Albert Popp and Bruce Lipton have initiated a revolution in our understanding of genetics. They've demonstrated the role of biophotons that, when coupled with human intention, activate our DNA's potential. The evolutionary potential of the divine human already exists within our genetic codons. It is just waiting to be turned on!

It may be that biophotons are directly receptive to the cosmic rays and gamma rays now pouring in from the galactic center on the incoming galactic superwave. This could well be what the descent of supramental consciousness is about. If so, we may soon experience a tremendous acceleration in biological evolution.

The metaphor of the butterfly may be relevant here. The caterpillar can jump up and down all it wants but it will not become a butterfly until the timing is right. When the timing is right no level of resistance will halt the transformation. This cosmic timing is what Year Zero is referring to. It is a reset in consciousness resulting in the emergence of the butterfly human!

As I shared earlier in this book, a profusion of crop circles are appearing around the world. Some kind of higher intelligence is interacting with humanity and the consciousness of an emerging Gaia field. One such pattern that showed up in the Netherlands a few years ago is the human butterfly. If the caterpillar symbolizes the human human, the butterfly represents the divine human! It appears we are now in the early stages of this metamorphosis, as reflected in this crop circle!

In recent years, my own explorations of metamorphosis have led me to working with a morphogenetic field of divine light known as Ilahinoor — rooted in the mystery school traditions of ancient Egypt.

Ilahinoor is an ancient and universal morphogenetic field designed to prepare the physical body for merging with the light body. This merging is facilitated by creating an energetic connection to various centers in the brain, which helps to awaken a cellular intelligence within the body. In response to higher vibrational fields of incoming cosmic light, deep subconscious fears and traumas are released from cellular memory, allowing us to experience and express the unified consciousness of our multidimensional self.

It is a simple, powerful, and effective means for healing the body, clearing emotional trauma, and manifesting our divine potential in these times of planetary birthing. Ilahinoor connects us with a morphogenetic field of vast wisdom and love — the realm of our souls. It helps to ground this supramental frequency of light into our physical bodies and daily lives. Thousands of people have been using these techniques in recent years with extraordinary results, as I share in my book — *Ilahinoor: Awakening the Divine Human.*

The higher we reach, the lower we must embrace. The task of supramentalizing our bodies is not easy. Not everyone will succeed, at least not in the early stages of the Shift. For those whose souls have chosen this path, it is a journey deep into our subconscious where we integrate the lower self with the higher self. This journey requires a willingness to die to our conditioned personality — our ideas, illusions, and judgments about who we think we are.

We can expect to go through a dark night of the soul — not unlike the caterpillar entering its cocoon — where we feel profoundly disconnected to everything we have identified with in the past. We may experience restlessness, anguish, and discontent. We may no longer feel our connection with spirit. We may go through a period of intense emotional clearing as the light of our divine self illuminates the hidden recesses of our subconscious mind.

While our physical bodies undergo a frequency adjustment, we may experience symptoms such as heart palpitations, headaches, body aches, insomnia, fevers, nervous system disorders, and kundalini activations. Our memory may suffer, and we may feel like we are losing our minds. Our sense of time may alter, prompting us to feel as though we are living in two or more realities at once.

Despite all this, there is a harmonizing principle within this supramental force that will make our work easier. Eventually, in the fullness of time, the butterfly will emerge. If even a few succeed in supramentalizing their bodies, the ripple effects on humanity will be profound. For all of us who have chosen to partake in this Shift no greater adventure can be imagined. I would rather be here on this planet at this time than anywhere else in all the dimensions of creation!

CHAPTER 42
RELATIONSHIFT

One area we are beginning to experience this Shift is in the arena of human relationship. This applies to all forms of relationships, yet I would like to share some insights here particularly regarding romantic partnerships and sexuality, since this is an issue that almost everyone these days seems to be dealing with.

Relationship paradigms are changing. When I am in touch with my essence, I realize I am no longer seeking to be with somebody for any of the old reasons — to make me feel happy, fulfilled, or alive — but someone who can partner with me in the task of creating a new world together. I don't need to possess, own, or control another person. I regard them as free to live their divine destiny as completely as possible. I don't even need to define what the relationship is about. I simply bless and honor the soul connection that exists and let it unfold and change with the seasons.

This isn't always easy. We are caught between the old and the new. We are constantly dealing with old fears, insecurities, jealousies, and dysfunctional patterns. We grapple with our own expectations of ourselves and our partners, as well as the expectations from the world around us. We haven't truly created viable new models yet. It could be that there aren't any defined models to create anyway, just a process of listening to our hearts in each moment and trusting an evolutionary impulse that doesn't always make sense to our conditioned minds.

When I reflect on what kind of new paradigms we can create together, I think of the futuristic novel by Thea Alexander titled *2150 AD*. It speaks of an entire society built on a sense of vibrational resonance rather than the guilt-ridden, convoluted, social conditioning that drives us today.

There is the recognition that we travel in soul groups and we find our sense of belongingness within this pod consciousness. We are not looking for a partner to magically fulfill every need we have. We instead ask ourselves what we can give to, rather than receive from each other. In this context, relationships and sex are a means for deeper communion with the whole. The more unconditionally loving we are, the more fulfilled we each can be within this pod consciousness.

In the society of 2150 AD, people have learned to move beyond the competitive model based on survival of self into a pod consciousness — where they derive their sense of identity from the entire soul group. There is the recognition that our highest joy is to support each other's highest joy. Nobody owns or controls anybody else because other is not separate from self. There is no need for jealousy or possessiveness because there is no sense of ownership. And children are raised by the entire community, honoring each other for their unique contribution to the whole.

Is this a realistic model for us today?

I believe that the human species is in the midst of making an evolutionary leap from the animal human to the divine human. The paradigm shift that is taking place in relationships today is a response to this evolutionary impulse. The more attuned we are to this impulse, the more we will be called to brave these uncharted waters. Each person's journey of relationship can be absolutely unique. It may not even be about creating an alternative model of relationship but about learning what it means to live from our hearts in each moment and to trust our unique process based on our own attunement to the emerging divine human. It is a process of maturity that grows from the inside out.

I attended a lecture by shamanic astrologer Daniel Giamario many years ago in Hawaii. Giamario lectured on the shift from a Cancer-Capricorn axis of relationship to a Leo-Aquarian axis of relationship. This type of relationship paradigm shift is what characterized the Sixties revolution.

The energetic signature of the astrological sign of Cancer concerns home and family; Capricorn, its complement, concerns security and stability. This was the ideal of the nuclear family — the task of raising children took precedence over everything else. Relationships were meant to last a lifetime. The couple referred to each other as their other half. Each person's role in the family was well defined.

Then came the shift to the Leo-Aquarius paradigm. Leo is about freedom and individuality, breaking free of personal limits. Aquarius is about expanding boundaries, moving beyond the nuclear family to the global family. This was the "free love" generation, which eventually settled into the multiplicity of relational forms we encounter today.

The purpose of relationship here was no longer to simply raise a family but to discover personal fulfillment. We began to find that it wasn't quite so easy to balance personal fulfillment and family obligation. Permanent monogamous relationships became increasingly rare. Divorce and serial monogamy became common, as did experimentation with various levels of polygamy.

An intense need for self-understanding in a social context emerged as we moved closer to the Aquarian age, which meanwhile gave rise to an explosion of the new consciousness paradigm. A multitude of psychotherapeutic models and self-help practices surfaced to help people and their relationships.

Much of the struggle and heartache in human relationships seems to be based on the apparent clash between these two relationship models. But it seems to me there is a third model emerging now — a Virgo-Pisces paradigm — which may hold the key to balance and wholeness.

The characteristics of Virgo include selflessness and service, while Pisces is about spiritual transcendence and soul connection. As we step into a new age of human relationship

209

based on an evolutionary shift into the divine human, we are beginning to see a new paradigm of relationship. I call this new paradigm relationshift.

This is what I feel the futuristic society of 2150 AD was practicing. It is an ideal that we can begin to model today. Such relationships are built on a foundation of soul resonance. They are based on the recognition that we travel in soul groups rather than as separate individuals. The more connected we are with our own soul resonance, the easier we recognize others in our soul pods. As we connect with our soul pods, our growth and value comes from how much we can give to each other. There are no fixed forms, just the movement of life in accordance with the constant guidance of our souls.

Inevitably, as long as subconscious tendencies of ancient trauma remain within our cellular bodies, we might find ourselves experiencing moments of jealousy, possessiveness, fear, loss and pain. This becomes the gateway to our own spiritual enlightenment as we recognize that there is nothing we can ever lose once we are connected with the higher resonance of the soul pod. The outer beloved becomes a symbol of our inner beloved. Feelings of separation give way to the experience of soul unity and unconditional love.

How many of us have craved for our ultimate soul mates, or twin flames, as the case may be? How many of us have refrained from fully engaging in the relationships of our lives in this elusive search for the ultimate missing half of our soul that would make our lives forever complete?

Perhaps some of us have been fortunate enough to find this 'perfect' partner, or have simply been willing to live fully in the present moment. But for those of us who have lost ourselves in elusive fantasies, could we be looking for our own inner beloveds reflected in each face that we encounter in the world around us?

Once we come into this recognition, we are ready to move into a third phase of relationship. This is where dependencies and co-dependencies end, and relationshift begins. This is where each relationship in our lives, no matter what form, becomes a gateway to the discovery of our own true face.

The three models of relationship I have described above are not mutually exclusive. I may choose to be in relationshift with someone that lasts a lifetime; or, I may choose never to commit to any one person. I may choose to have children or not. I may be in an open relationship, a homosexual relationship, endure a string of partners, or remain celibate. The forms are not important. What matters is that we are open to the flow of love through our souls, finding its most natural expression with each person we meet without judgments or fear.

As we move into this time of the Great Shift, our human biology is undergoing a change as well. As we become adept at shifting through dimensions, we may not need to bring children into this world through sexual reproduction.

Sexuality would no longer then be coupled with an overriding instinct for species propagation with all the darker elements that tend to go with it in these times. It may be easier then to simply live from a place of soul contact — without walls and without fears — as in Thea Alexander's utopian vision of the future.

Going through these stages has been a long journey of discovery for me. It has been sometimes immensely fulfilling and sometimes sad and lonely, even heartbreaking. My close friends sometimes tell me I am not always good with receiving. Maybe my childhood patterns and conditioning propel me to downplay my own needs for personal happiness. I, in turn, justify my stance that planetary wellbeing is more important. This can even lead to a bit of a martyr complex — other people's needs consistently become more important than my own.

However, I am beginning to realize that the two cannot be separated anymore. Service does not mean martyrdom. As a Pisces my tendency is to use transcendence as an escape from engaging fully in my own humanness. I have used my willingness to serve the planet as a distraction from being aware of my own needs and feelings. Oftentimes I have neglected my physical body or pushed myself too hard while denying sensory pleasures or joy.

The Virgo-Pisces axis is about loving my neighbor as myself. If I cannot love my partner, I cannot love myself. But equally, if I do not truly love myself, I cannot love my partner either. In relationshift, we learn to receive as much as to give. We learn to experience personal fulfillment and joy as much as to hold space for another's. Giving and receiving, service and joy become two faces of the same experience of oneness.

To love unconditionally we make no distinction between self, partner, and planet, for we are all part of the web of life together. The experience of joy is the highest expression of our divinity. Relationshift is a gateway towards this dance of the soul.

A new paradigm based on unconditional pod consciousness cannot be imposed from the outside. It is something that must emerge from within. Sometimes this can be painful and confusing. It is a process of breaking out of our old emotional and societal conditioning, and requires a great deal of awareness, self-respect, and trust in the process of life.

We will likely make many mistakes along the way. We will vacillate between the new paradigm and the old. In learning to love another more fully, we are learning to love ourselves more fully. We are unraveling millennia of deep conditioning. Ultimately, it is only as we acknowledge our own deepest soul as the inner beloved we have been seeking all along, that we can truly love our partners, our children, our neighbors, our friends, and ultimately the entire human and planetary family.

CHAPTER 43
REINVENTING HISTORY

"He who controls the past controls the future".

Who pulls the strings behind world events? How much do we as a people control our own lives and destinies? Why are those whom we have elected to govern us often the first to betray our basic rights to life, livelihood, freedom and happiness? Is there a hidden agenda at work? Is it possible to change this?

Some time ago I came across the work of Russian mathematician, Anatoly Fomenko, who wrote a series of books trying to prove that an entire thousand year period of human history had been completely fabricated, and never actually happened. As bizarre as this might sound, he makes a very convincing case out of this.[1]

Whatever the truth of this claim might be, the history of the human race is filled with war, conquests, lies, and atrocities. The winners came away with land, resources and booty, but more importantly, they got to record how future generations would remember the story. The winning side was usually portrayed as the paragon of honor, bravery and righteousness, while the losing side was always demonized in order to justify villainous acts of greed, treachery and genocide. This is the basis for patriotism, or loyalty to one's state, and nationalism, the assumption that one's own country or tribe is best, and always right.

Empires have been built on these stories. Could the Greek empire have survived without Homer to sing the great and glorious praises of heroes like Achilles and Odysseus as they went about their business of slaughter and pillage? How would history read today if Troy had managed to resist the Greeks? What if the Confederates had won the American Civil War? Or if Germany had emerged victorious in World War?[2]

I recently listened to a speech given by Adolf Hitler in 1939, just before the start of the Second World War. He is passionately calling for peace and sovereignty for all people in Europe, asking them to stand together and liberate themselves from the control of the Rothschild banking system, which he blames for the creation of the First World War, as well as for the crippling sanctions on Germany that followed. He accuses Churchill for wishing to launch a war that would plunge the entire continent into darkness on the behest of his masters, this same banking system.[3]

We know today that the Allies won that war. We know that the banking cartels were funding and arming both sides of that war, at enormous profit to their own cause, and enormous misery to the millions who lost their homes, lives and sanity. We know that World War 1 was manufactured out of total lies for the financial gain of JP Morgan. We also know that the power of this banking cartel has not ended, and continues to fuel the flames of war and terror throughout the world.

Could it be that Hitler's assault on the Jews was not so much about targeting individuals but an attempt to overthrow the Rothschild/Morgan banking cartel itself? And could it be that this seven-headed Hydra has been surfacing again and again around the world to continue its assault on peace, security and freedom?

The names have changed, and the players have changed, but the game continues. In the days of maritime piracy, pirate ships would often fly a 'friendly' flag as they got close to their intended target, unfurling their true colors only as they got ready to ram, board and slaughter their prey.

We know today that every major war of the twentieth century was initiated by a 'false flag' attack. Many recognize that our current 'war on terror' was also initiated the same way. The strategy of demonizing the enemy to gain support for an assault, cleverly staging a false flag event, and then sending in troops and bombs to 'liberate' the innocent victims of this demon dictator, continues to be as effective today as it ever was.

How much do we really know about Adolf Hitler, other than what is known from the winner's point of view? I use Hitler as an example because he has probably been demonized more than anybody else in history, with the possible exception of Attila the Hun or Genghis Khan. But there are other statesmen and political leaders in the Middle East, Latin America, Africa and the Far East, who have been equally vilified in order to justify their removal from power, whose biggest crime has been simply to resist this agenda for global hegemony.

Who are the real villains here? Is there such a thing as truth in a world where the media has become the propaganda arm for the winning side, in this case the banking cartels, to stage a war of perpetual terror and destruction that threatens the survival of the entire planet today?

I am not suggesting that Hitler was a saint. He was tormented by his own demons, as was Churchill, and the war that was unleashed in response created scars that will perhaps never completely heal. But how would history judge him today if he had not been pushed back against the wall, and allowed to continue the economic reforms that brought Germany back from the brink of total collapse after their defeat in World War 1? Was there perhaps a hidden evil never acknowledged as the winning side promptly rewrote the history of those times?

There is no person, no nation, and no form of government that is absolutely good or absolutely evil in this world. Yet there are certain institutions and policies that, unless they operate within a system of checks and balances, can threaten the well being of our entire civilization. Countless books have been written about this, and it is not my role here to elaborate on this in detail, but it is important to look at why the world is still at war when most people only want peace, why so many are destitute, desperate and dying when a privileged few have access to unimaginable resources used only to further subjugate and oppress the many.

I believe that people are basically good at heart, and do not wish to kill, maim and terrorize each other. I also know that as long as we remain in a state of ignorance, deprivation and fear, we can be easily duped and deflected into carrying out whatever agendas are demanded of us by those in power.

I do believe that there is a hidden agenda at work, and that this is sourced in our current banking system. Many have referred to this hidden hand as the Cabal, or the Shadow Government, or the Deep State, a trans-national group of bankers, industrialists, state agencies and political leaders who have been following an age old script for absolute financial control and global domination.

As I write this chapter, there are proxy wars being fought in Syria and other nations in the Middle East, bringing us perilously close to yet another global war. Can we begin to see this not as an isolated instance but as a predictable move in a deplorable War of Thrones?

Those who have learned to manipulate history know how to control the future. We have given our power to a deliberately misleading mainstream media that serves a dark agenda simply because they are financially beholden to this global scam network. Those who resist this agenda are often scornfully labeled as conspiracy theorists, crackpots and nutcases. The tide is rising, however, and as more and more of us begin to wake up, look behind the smokescreen, and question the conditioned reality of our present-day matrix, we might truly have a chance at creating lasting peace on Earth.

CHAPTER 44
A UNIFIED WORLD
ORDER

It is no measure of health to be well adjusted to a profoundly sick society.

J. Krishnamurti

Back in my college days I was actively involved in issues of peace and social justice. I was enthused with the idealistic hope that all it took was a few dedicated visionaries to change the world. I was not always able to sustain this hope. On the one hand I believed we were creator gods capable of anything; on the other, I struggled to find my own balance in a world spiraling out of control.

Somewhere inside I knew we were capable of making the greatest evolutionary leap this world had ever seen. At the same time, I felt a profound despair and intense rage toward a species that was so foolishly and aggressively ripping the delicate web of planetary life to shreds. There were even times I wished for the end of all humans on Earth so that nature could survive, heal and thrive.

Eventually I fixed my hopes on an upcoming shift in consciousness, much like various others who had pinned their hopes on a religious savior or extra-terrestrial intervention. The global situation appeared too desperate and hopeless. We were spiraling too close to the brink of extinction. I had lost faith in our ability to pull through on our own.

Over the years I have found my faith again. But it is not a blind faith. More than ever before I believe in a collective shift in consciousness. But while I do feel this shift is being orchestrated from a higher level of consciousness, it also requires our active focus and conscious intention.

We are co-creators in an emerging new reality. But we cannot be creators and victims at the same time. We cannot remain passive observers. Are we willing to envision a courageous new world and fully commit our resources to empower this vision? Are we willing to put our feet on the ground so we can walk tall and strong towards this promised land?

This question becomes especially relevant as we consider our economic and political systems. It is clear that if we are to survive on this planet and evolve beyond the roots of greed, corruption, competition, and manipulation, our current systems need to change. Change begins with understanding. If we wish to create a new world, then we must open our eyes to the big picture and anchor this vision into the physical world. As Goethe stated so profoundly centuries ago, *"No one is more hopelessly ensnared than those who falsely believe they are free."*

Our current economic system is based on money. The paper money system, rooted in the fractional reserve system, is directly linked to debt. Debt is linked with scarcity. Scarcity is a requirement for the creation of profits. And profit, as far as our capitalist economy is concerned, is the only incentive capable of promoting growth, providing employment, enhancing creativity, and sustaining our way of life in an increasingly debt-ridden world.

Banks, businesses, and corporations have become the guardians of this system — euphemistically referred to as the free market economy. The profit motive inevitably leads to greed, impoverishment, corruption, callousness, environmental degradation, social injustice, and war. Resistance to the system is invariably discouraged or crushed by governmental bureau-cracies, media control, and military might.

Social and environmental costs are ignored. The rich become richer, the poor become poorer, and species continue to become extinct.

We examined earlier the potential of solar, wind, tidal, and geothermal power to replace fossil fuels, as well as the emergence of free energy technologies. We already possess ingenious technologies to clean up our planet and restore her biodiversity. We have the resources to easily feed, clothe, house, and educate every human being on Earth. We could live in abundance, harmony, and peace at a tiny fraction of the cost of maintaining our global military systems.

So why isn't this happening?

In his book Confessions of an Economic Hit Man, author John Perkins offers shocking revelations about how our current banking system, corporate interests, and military interventions play a part in undermining national as well as personal freedoms around the world.

He speaks of a shadow government operating behind the scenes, sometimes referred to as the oligarchy, cabal or the new world order, which has been in the business of toppling nations and economies for a long time now, whether in Central or South America, Southeast Asia, Africa or the Middle East, and recently extending into Europe and the US as well.

This is usually initiated through bribery and debt creation through economic hitmen and jackals, operating through multi-national corporations supported by agencies such as the CIA. If these efforts failed the next step was to either plot an assassination or stage a military invasion and takeover with the intention of installing a puppet regime more in alignment with US corporate interests.

The US has invaded 70 countries since its inception. It is involved in armed conflicts within 134 countries right now. Thirty-seven countries around the world have experienced direct military intervention or CIA-initiated coups since World War 2, with an estimated 20 or 30 million people killed as a result. When Americans naïvely question why it is that so many around the world 'hate America', is it possible these are some of the reasons why?

Perkins is speaking specifically of the USA, but the same strategies have also been used by other imperialist states around the world. He unpacks how our global political systems are a direct extension of economic and military-industrial interests.[1]

Likewise, a wide range of alternative energy technologies and health breakthroughs have been systematically ignored or suppressed because they cannot be controlled for profit. Meanwhile, our dependence on fossil fuels, nuclear energy, and polluting technologies has brought our planet to the edge of irreversible climate change and environmental disaster.

Inventors who have developed working prototypes of free energy technologies are reluctant to put them on the market because they fear the consequences to themselves or to their families. I know some of them personally, and I have glimpsed how their discoveries could change the world. If Earth is to survive, and if our civilization is to rise to its highest human potential, we must make room for these technologies.

For the most part, our political systems have their own vested interests in perpetuating the existing state of affairs. As the world's reserves dwindle, wars are fought in the name of oil, uranium, and other natural resources. The world's last preserves of wilderness are being drilled or fracked for oil, or destroyed for lumber and for raising cattle. Wars are being fought over pipelines. Millions of people are being relocated to create hydroelectric projects to supply our needs for power. Is all this necessary?

It appears that there are two remaining superpowers in the world today. The first is the United States of America, secretly ruled by the banking elite and transnational corporate interests. The other is our unified social consciousness. This collective soul force, as harnessed by Mahatma Gandhi and Martin Luther King, can redesign the world towards sovereignty and freedom.

The Zeitgeist Movement, advocated by filmmaker Peter Joseph, and The Venus Project, inspired by futurist Jacque Fresco and Roxanne Meadows, are beautiful examples of how this social consciousness is emerging in the world today. In his films *Zeitgeist: Addendum* and *Zeitgeist: Moving Forward*, Joseph calls for a resource-based economy rather than a monetary system; a model based on abundance and equality rather than scarcity and control.

Our current system, driven by profits, is based on an Orwellian double-speak, where abundance is scarcity, and scarcity is abundance. In other words, when supply is limited, the demand goes up and profits increase. Shortages and chaos are manipulated for profits and are built into the system. Those already rich get richer, and the poor get poorer.

If scarcity is directly proportional to profits, then sustainability and abundance are the enemies of a profit-based system. An abundance of basic resources, such as food, water, medicine, alternate or free energy technologies, when freely available to everyone, would inevitably lead to loss of corporate control and profits, and therefore cannot be tolerated in a 'growth' economy. Maximizing profits, regardless of social and environmental costs, is what free trade and globalization are about.

As long as we follow our standard monetary system driven by profits, it is impossible to create a world without poverty, crime, pollution, unemployment, anxiety, and war. As long as profit means competition, it is impossible to behave ethically or compassionately. As long as we perceive the Earth as simply a resource to be used up, we head toward environmental catastrophe. As long as we remain enslaved to such a system, we cannot hope to unleash our innate creativity or to solve the complex problems we face on Earth today.

Whether capitalism or socialism or communism or globalism, all these systems are based on the illusion of unlimited resources and on the idea that profits and money have their own intrinsic value, which must be pursued at any cost. My value as a human being is measured in how much money I control.

"Make no mistake," says Joseph, "the greatest destroyer of ecology, the greatest source of waste and depletion and pollution, the greatest purveyor of violence, war, crime, poverty, animal abuse, and inhumanity, the greatest generator of social and personal neurosis, mental disorders, depression, anxiety, not to mention the greatest source of social paralysis stopping us from moving on to new methodology, personal health, global sustainability and progress on this planet is not some corrupt government or legislation, not some

rogue corporation or banking cartel, not some flaw of human nature, and not some secret, hidden cabal that controls the world: it is in fact the socio-economic system itself at its very foundation."

Sooner or later, an economic system that ignores ecological sustainability and focuses only on short-term profits is bound to collapse upon itself, taking the rest of the planet along.

On the other hand, Fresco's resource-based economy focuses on the availability and distribution of the Earth's resources — our common heritage.

There is no reason for anyone to experience hunger, exploitation, terrorism, or war. There is enough food, water, and energy to abundantly meet the basic needs of every single human being on Earth. It is a matter of redirecting our priorities.

If we choose to free ourselves from a money-based system, all existing resources could be used to serve the whole rather than the few. It would be based on creativity and the use of synergistic technologies to enhance human life. This would ultimately move us into the realm of the divine human.

Can we create a system with a standard of living so high that we do not need money for survival? Can we design a society that would eliminate boring and repetitive jobs, eliminate poisons in our foods, and develop clean and efficient energies — a society truly concerned with the well-being of people and the planet?

If each of us were assured of all of our basic human needs, we could express our unique creative gifts to serve the whole. Creativity, altruism, and inspiration would dwarf profit-motivated growth. We would develop systems that enhance rather than exploit creation; serve rather than destroy life. And we would quickly focus our vast collective intention on averting the looming, massive environmental cataclysms. This is what the Venus Project hopes to achieve.

The most important priority, Joseph summarizes, is the intelligent management of the Earth's resources. He advocates for all natural resources on the planet becoming the common heritage of all people. It is time to think outside the box. Instead of weapons of mass destruction, it is time to unleash a weapon much more powerful — weapons of mass creation!

Joseph's documentary films develop these themes with inspired conviction. They have been shown in movie theaters around the world, and are available on the Internet for free. The ending of *Zeitgeist: Moving Forward* is especially inspiring. It's reminiscent of the Matrix movies and well worth watching.[2]

There are many viable alternatives to the self-imposed enslavement we have chosen to experience on the planet today. Movies such as Avatar and The Green Beautiful provide moving examples of what a new social and economic system might look like. Visionaries such as Barbara Marx Hubbard, Peter Russell, Makasha Roske, Soleira Green, and Daniel Pinchbeck have also created social and informational networks to facilitate a unified world order.

With all the fears and doomsday scenarios bombarding us today, the worst thing that can happen is that nothing changes. Unless we choose to turn things around soon, we will inevitably strangle in the rapidly tightening noose of our own extinction.

A wide range of contemporary economists, futurists, and social scientists predict an imminent, worldwide collapse of our monetary system. It has, built within itself, the seeds of its own demise. It is no longer a question of whether the system will collapse, but when this would happen, and what would be the repercussions of this.

Frightening as this all may sound, this could ultimately be the best news we could hope for. Pluto is the planet of deep transformation — death followed by renewal and rebirth. Pluto recently entered into Capricorn, which represents our planetary systems and social paradigms. Meanwhile, Uranus, the planet of revolutionary change, has moved into Aries, where unseen fires can now be released. All that is hidden must inevitably come to light.

Our world is terribly out of balance, and it is likely that a collapse of the entire system could happen very quickly and suddenly. Are we ready for this? Is there something that can replace our current system? Can we make a smooth transition into a caring, sustainable, resource-based economy? Several nations in Europe have been experimenting with a basic income for all its citizens, allowing them to devote their lives to

223

creativity rather than survival. Perhaps this is a good place to start.

A new dawn is coming. As we reflect on the global shift resulting from the evolutionary forces moving through the Earth today, we cannot ignore our social, economic, and political realities. New systems and paradigms, more responsive to the call of a unified social consciousness, must replace the old. We can neither hide our heads in the clouds nor bury them in the sand. Political action without spiritual vision is not sufficiently inspired. Spiritual awareness without socio-economic vision is not sufficiently grounded. Together, using the power of creative intention along with inspired action, we can create heaven on Earth.

CHAPTER 45
JOURNEY TO PERU

In late 2010, I went to Peru with a small group of people. We travelled through various sacred sites with the intention of anchoring light deeper into the collective consciousness of humanity and Earth.

The final week of the journey was spent in the deep jungles of the Amazon with an Asháninka shaman. We experienced three sessions as a group with the powerful plant medicine known as Ayahuasca. Then I continued on my own for another four days. What I experienced during that time has significantly changed my perceptions of reality and my understandings of the Shift that we are currently experiencing on Earth.

The Amazonian shamans speak about three worlds that are constantly interacting with one another. Hanaq pacha is the upper world, the world where our higher selves reside, the angelic world, the realm of the condor. Kay pacha is the middle world, the world of our human selves, the animal world, the realm of the jaguar. Ukhu pacha is the lower world, the world of subconscious instinct, the reptilian brain, the realm of the anaconda.

I began to explore these three realms by creating bridges between them. The beginning of each journey was dominated by snakes and dragons, and images from the deep subconscious.

The images reflected the separation and fears held within the instinctive centers of the brain.

Over time, as I allowed these images to shift and change with awareness rather than resistance, the power of this instinctive world began to reveal itself. Rather than being a repository for subconscious conditioning and fear, the instinctive self became a gateway for contact with the higher self. The anaconda joined with the condor. The personal self emptied out. Vast beings of light anchored through my body. I felt the llahinoor fields of light merging with the soul and body of Mother Earth, moving through the nature kingdoms, preparing Earth for the dimensional shift to come.

I felt the presence of Shiva incarnating into various Earth dimensions. A vast power of transformation, I felt he could have uprooted everything that human society had ever created in a relentless urge to break the spell of third-dimensional illusion. I felt the energies of Divine Mother merging with Shiva. *"Deep transformation, no destruction"* was the mantra that echoed through my expanded consciousness, setting an intention for planetary evolution that harmonized with the supramental descent of cosmic light.

This process went on for seven days. Bolts of lightning rippled through my body in response to a secret intention within the spirit of the Earth. Again and again I doubted that my body could take much more. And yet, each time, as my shaman held out his cup, I did not hesitate to drink fully. I knew this was a pact I had made with Mother Earth to serve her in this time of birthing.

I realized something through it all. There came a time when the instinctive fears emptied out, when a great power opened itself within the hidden centers of my being. The elemental dragons, who are the guardians and protectors of Earth Mother, came alive. The creative power of the lower worlds could merge now with the forces of love actively held within the upper worlds, sustaining a balanced frequency of light strong enough to liberate the illumined potential of the divine human.

I realized this was a cellular process. An old program of subconscious fear, separation, and duality was ending while a new program of multidimensional presence and physical immortality was being activated on a cellular level. I was guided by the resonant totem of the anaconda, which represents the power of divine potential, and is the guardian of alchemical transformation from physical matter into true matter.

As the process continued and deepened I could no longer function in linear time, which is the realm of the thinking mind, and was therefore no longer subject to the limits of human conditioning. It was only afterwards, as I returned home from the jungle, that I realized the depths of the transformation that had taken place. As I wrote in my journal afterwards:

"The cosmic fields of light feels more expanded and yet more anchored within the kingdoms of the Earth. My physical body feels younger and more vital. I can hold enormous fields of light within my expanded self. I feel my body merging into silent union with all kingdoms of life.

"In a place beyond the realm of the thinking mind, I know and understand that the birthing of a new consciousness is at hand. The separate human identity is giving way to an expanded planetary identity. It is Gaia as a planetary being who is awakening now, not just the human species. Throughout the galaxies, all creation is undergoing a basic shift in vibrational frequency. As a consequence of this, for the first time in 13.6 billion years, creator consciousness is slipping through the veils of time and space into the heart of creation, waking up within the dream, fully present in the world of matter.

"Herein lies the power of my own creative energy. In this state of consciousness I am no longer identified as Kiara. I am no longer identified as human. I am creator consciousness at play in the field of matter, awake simultaneously in the three worlds of creation. I am not separate from the Earth. I am not separate from the Skies. I am an open door between Earth and Sky through which supramental light can flow!"

CHAPTER 46
EVOLUTIONARY PUZZLE
PIECES

L et's bring together some of the puzzle pieces we have examined so far, before moving on to one final piece in this evolutionary jigsaw puzzle, which points to a possible mechanism for how an evolutionary shift could happen.

Many advanced souls have chosen to incarnate on this planet now, simply because these are the times when a massive transformation of planetary realities is at hand.

This transformation is directly linked with changes taking place in the Earth's magnetic field. The magnetic fields of the Earth are a template for consciousness, and reflect the long trajectory of biological evolution through millions of years of Earth's history.

The Earth is a complex biological entity and like all forms of life, has her own cycles of expansion, contraction, involution and evolution.

Many scientists today have come to realize that the universe is a multidimensional field of electrical energies, constantly exchanging information with other galaxies, suns and planets, the physical body of which is simply an extension of the primal consciousness that permeates all things.

Our universe did not emerge out of nothing in one Big Bang, but is continually experiencing new waves of creation as primal consciousness expands out in evolutionary waves from the centers of galaxies.

These waves of evolution are conscious and intelligent, carrying within them the template of new creation, new seeds of life travelling out on cosmic waves of light throughout galactic, solar and planetary systems.

The ancient seers of India understood these galactic waves as the force initiating the four *yugas,* and understood that they passed through our solar and planetary system in regular and predictable cycles.

As scientists are discovering today, these same galactic waves correlate with long-range cycles of solar activity as well as with magnetic reversals upon Earth.

A magnetic reversal refers to a periodic event when the overall magnetic field of the Earth begins to weaken very quickly over a short period of time, followed by the poles flipping towards a new axis. As the new polarity establishes itself, the magnetic field strength gets recharged.

The periodicity of these interconnected cycles seem to follow the Milankovitch cycle of roughly 11,500 years. Many researchers are coming to recognize that it is these cycles that drive biological evolution on Gaia.

The most recent magnetic reversal on Earth took place almost exactly 11,500 years ago, and observations from NASA, NOAA, ESA and other research facilities indicate that we are in the midst of another magnetic reversal right now.

Observations of earlier cycles of magnetic reversal indicate that, once started, the entire process of magnetic reversal could complete itself within a single human lifetime. The current cycle of reversal seems to have begun in the early 1800s, starting rather gradually, and moving rapidly towards an exponential collapse, or zero point.

Research indicates that the last phase of a magnetic reversal is generally a dramatic event, where the Earth's magnetic field disappears almost completely, followed either by a reversal, excursion, or a 90-degree phase shift.

This last phase of collapse, referred to in indigenous prophecy as the three days of darkness, has been awaited by many cultures as a time of purification and transformation. Since the magnetic field of the Earth is a template for the collective memories of Gaia, a collapse of these fields could portend a radical shift in the collective identity of Gaia, in which we as a human species would play a significant part.

The repolarization of these fields then would signify a re-set within Gaia's evolutionary program. Indeed, according to some biologists and science researchers today, this is exactly how the evolution of species takes place. During times of magnetic reversal, Gaia experiences a quantum shift within her biological fields, whereby entire species go extinct and new species emerge.

This happens because the Earth's magnetic field normally serves as a shield against solar and cosmic radiation. When the shield is down we are susceptible to massive amounts of solar radiation and cosmic radioactivity, both of which are capable of drastically affecting biological life and climate patterns on Earth.

Now, perhaps for the first time in Gaia's history, there is a species of life on this planet that has become self-reflective, capable of taking the forces of evolution into our own hands, and driving it forward consciously. Our species serves as an intergalactic nervous system linking the vast network of life on this planet to endless fields of life across the physical and spiritual universes.

The mighty forces that drive evolution also move through our human systems. They operate far beyond conscious levels of awareness, responding to a supramental field that is linked with deep levels of the collective human soul.

Thus, as the Earth's magnetic field collapses and reverses, immense waves of evolutionary energy move through our collective human matrix, inviting Gaia to move into the next phase of her own evolutionary journey.

Solar and cosmic radiation moving through the Earth's atmosphere is capable of causing massive extinctions of species during this time due to biological mutation. But this same radioactivity is also capable of generating new biological life forms.

Our current human species, anchored within the matrix of separation and duality, could well be due for extinction. However, a new species, which the Mayas referred to as homo luminous, is emerging from within our current species. In this lies the hope for a New Earth that has been predicted by so many shamans, mystics and seers for centuries past.

CHAPTER 47
DMT, DOORKEEPER
BETWEEN WORLDS

We have established by now that the shift of consciousness predicted by many during these times is directly linked with the collapse and re-orientation of Earth's magnetic field. The collapse of the Earth's magnetic field has been proceeding at exponential speeds over the past 200 years or so.

The current average field strength is about 80%, although significantly lower or higher in certain parts of the world. I don't expect to see a final collapse of the fields until we are down to at least 15 or 20%. But the question remains, at the current rapid and exponential rate of collapse, how far are we from this occurrence? How close are we to the prophesied three days of darkness?

Barry Martin Snyder, author of the *Birthing the Luminous Self* trilogy, extrapolates from various charts and patterns to pinpoint the time of magnetic collapse and reversal to around 2024 or 2025, although this could change based on human consciousness or unknown geological factors. Or perhaps, as John Casey indicates, this could happen shortly after our current 30 year Cold Sun phase ends around 2045 or 2046, as giant solar flares initiated by a massive barrage of incoming cosmic dust lead to huge surges in the solar wind, overwhelming and reversing an already weakening magnetic field on Earth.

Not only are the magnetic fields weakening, but the north and south poles of the Earth have been shifting towards new positions, also at an increasingly rapid speed. They are currently moving relatively independently of each other towards a point of convergence somewhere in Indonesia, with an opposite polarity emerging across the planet in the Caribbean, in the region of the Bermuda triangle. Thus rather than flipping 180 degrees all at once, the new magnetic axis could possibly line up at a 90 degree angle to the current polar axis, remaining there for some time before re-aligning with the original north-south axis.

What can we expect on Earth during this time? In the absence of the magnetic shielding provided by these fields, we would become susceptible to massive solar radiation as well as to cosmic radioactivity. What if there was an X-class solar flare during one of these periods of collapse, or even a relatively smaller M-class flare? What would happen to our satellites, power grids, and communication systems? What would happen to the sense of human identity that is normally anchored within our mental and emotional bodies? What would happen to our perceptions of reality? Would our bodies survive the immense amounts of radiation and radioactivity raining down from the skies?

I do not have the answers to these questions, but we seem to be moving towards an evolutionary crisis. During the conception of human life, millions of sperm cells are released and move towards the egg, but only one or two eventually make it to their destination. But conception could not happen without the support of these millions of sperm cells. Are we in a similar phase of planetary conception, where billions of human beings have taken birth on this planet in order to prepare the species for the next stage of evolution?

The science of sacred geometry has shown us that 90-degree phase shifts are the key to traversing between dimensions of time and space, sometimes referred to as timelines. As the magnetic polarity of the Earth re-orients towards a new polar axis at right angles to our current axis, it could open the doorway for a powerful shift for humanity and the Earth.

What do these scenarios mean for us? Can we continue to live out our petty human lives alternating between violent upheavals and quiet desperation, or is there another biological imperative that is coming to the forefront now? We have come to a fork in the road, and don't have much time to choose. What is being required of us in order to make this biological shift?

What will happen when the first signs of magnetic collapse take place? It is very likely that if this happens at the same time as a strong or even relatively moderate solar eruption, this could severely disrupt the Earth's power grids. Low orbit satellites would get fried, while power lines and transformers around the world would fail, perhaps for a period of months or years. There could also be a significant warming of the oceans on the side facing the Sun, leading to severe storms and hurricanes, as well as massive earthquake and volcanic activity. Any of these scenarios could significantly set back our current human civilization.

This is the downside of a magnetic collapse. Meanwhile, there is the incredible evolutionary potential that has prompted seven billion human souls to enter this incarnational cycle. How do we use this opportunity as a species? It could well be that many will use this time to complete their incarnational contracts, while others will use it as an unprecedented opportunity to make an evolutionary leap.

Researchers have noted an inverse relationship between magnetic field strength and the production of a compound within the pineal gland known as DMT (di-methyl-tryptomine). As the magnetic fields of the Earth weaken, the production of DMT is enhanced. This is the substance in the brain, sometimes known as the 'spirit molecule', which allows us to transcend ordinary realities and experience multi-dimensional conscious-ness. It is the same ingredient found in plant medicines such as Ayahuasca, which opens doorways between the worlds.

What if the entire human species, as well as other mammals, began to experience a DMT spike during the magnetic collapse and reversal? Imagine the entire world on Ayahuasca during these three days of darkness! What does it take to consciously enter into the fields of matter and surrender to the evolutionary forces moving through us? Is the human species capable of making this shift, or are we too wounded and jaded from living at the edge of survival through so much of human history?

Like most quantum leaps, I suspect it takes only a few people initially to accomplish the genetic transformations required, and then, once complete, the rest can follow in future incarnational cycles. But for those who choose this work, it will be a massive undertaking, and not necessarily easy or painless. Are we willing to offer our vehicles to this transformational work?

There comes a time when the caterpillar can no longer go about its mundane existence crawling along the ground. The old structures of life are collapsing all around us. Political systems are tottering, economic systems are failing, social structures are in upheaval, our bodies and personal lives are falling apart.

Perhaps these old structures cannot be cosmetically fixed anymore. Perhaps it is our time now for entering the mysterious dark tunnel of a planetary birth, riding the evolutionary waves of new creation, and emerging as a human butterfly, the homo luminous.

"Man is a transitional being," said the great mystic seer Sri Aurobindo more than a hundred years ago. *"Evolution is not finished... As man emerged out of the animal, so out of man the superman emerges."*

Perhaps this is why many of us have chosen to be here at this time. A new age is at hand. The journey through the evolutionary marshes may be dark, difficult and unknown, but the map lies within our genes, and the way forward is joyous and sure.

CHAPTER 48
ELECTRIC UNIVERSE

There is an enormous paradigm shift taking place in scientific circles today about the nature and composition of the universe.

We are moving away from a gravitational model based on physical objects spinning around each other through vast interstellar distances to a plasma model where the entire universe is perceived as a single pulsating web of electric plasma, inherently inter-connected to everything else even beyond the boundaries of space and time.

Recall the ancient Indian metaphor of Indra's net, where the universe is perceived as a string of pearls, each pearl reflecting every other pearl, each pearl containing every other pearl, and each pearl essentially being every other pearl. Our physical universe is now being seen to behave far more in accordance with this model than our current scientific paradigm based on finite forces and linear motion.

Take for instance our own Sun. According to traditional science it is an immense ball of hydrogen and helium gases, producing heat through nuclear fusion, which then radiates out through the Solar System to warm up planetary bodies in proportion to their relative distance from the Sun. According to this theory there is a huge, but ultimately finite, amount of energy available to us, based on the life cycle of our Sun from white dwarf to red giant.

Likewise, according to this theory, the universe operates according to laws of entropy, beginning with the Big Bang 13.6 billion years ago, rapidly expanding out through an almost infinite stretch of intergalactic space, eventually slowing down, reversing direction, and collapsing back into a singularity of time and space. Although we are still billions of years away from this eventual point of demise, the universe is still finite according to this model.

Scientists have puzzled over why the universe seems to behave in certain predictable ways as far as these macrocosmic worlds are concerned, but works so differently in the microcosmic world of subatomic particles, where time and space take on an apparently random quality, and where the consciousness of the observer makes such a significant difference in how these particles behave.

Could it be that what we are looking for to unify these two worlds is not so much about discovering some elusive Theory of Everything, but shifting our perceptions slightly to understand what has always been right in front of our very noses?

Could it be that the laws of quantum physics and the laws of relativity are exactly the same laws, except that we just have not been able to perceive them rightly?

We have been so deeply conditioned by our physical senses to believe in ideas of shape, size and distance that it seems to me that we have lost our ability to observe the universal matrix underlying all time and space, all things in creation ranging from the largest super-galactic clusters to the smallest boson.

To our observable physical senses, it seems that gravitation is the strongest of the four basic forces. The entire fabric of the observable universe, including relativity, the speed of light, and the separation of objects in space and time, is based on our understanding of gravitational forces.

In our own Solar System, the planets seem to be held in orbits around the Sun by the forces of gravity, while the Sun is held in its 230 million year orbit around the galactic center through this same force. Our galaxy rotates around a super-galactic center, which rotates around its own galactic cluster, and so on, until the entire physical universe seems to be expanding out from its original singularity in time to the far extent of an almost infinite space, eventually returning back eons from now through the force of gravity.

But what if we were to perceive the universe not through the physical forces of gravity acting upon physical objects but though a vast network of invisible electromagnetic forces that weave all things together? Unlike the force of gravity, which diminishes proportionately the further two objects are separated from each other, electric forces maintain a connection between these same objects no matter how far they move apart in space or time. Across the vast distances of the cosmos, electromagnetic energy is a trillion times stronger than gravitational energy!

The Electric Universe theory, or Plasma Cosmology, is based on the idea that 99.999 % of the universe is plasma, and that plasma produces its own magnetic fields and electrical currents.[1] Wikipedia defines plasma as one of the four fundamental states of matter, the others being solid, liquid and gas. It is a quantum mix of negatively charged electrons and highly charged positive ions, created within hot gases or by subjecting gas to strong electromagnetic fields.

The Sun, stars, galaxies and even what appears to be physical matter, are primarily composed of plasma. Known to earlier alchemists as ether, it is the substrate of all things in existence.

Plasma cosmology takes away the requirement for a Big Bang, or any sort of beginning to space, time, matter or energy. It takes away the need for a single act of Creation. It takes away the need for dark matter or dark energy, black holes or neutron stars to explain away inconsistencies in the gravity theory of the universe. And it provides a meaningful context for string theory, which postulates that at the heart of all matter, is an eleven dimensional matrix of vibrating strings.

Physical interactions between objects on the macrocosmic level can be explained through plasma interactions appearing as electrical currents, while quantum interactions on the microcosmic level can equally be explained through plasma interactions between electrons, protons and a multitude of ever smaller sub-atomic particles. Ultimately, this theory explains why all matter, not just photons and subatomic particles, sometimes appear as particles and sometimes as waves.

It explains how human consciousness, when fixated in a physical universe of gravitational laws, seems to function through the limitations of the rational mind, through the laws of cause and effect, and within a matrix of time and space. In this state objects appear to be separate from each other, and the universe appears to operate through fixed laws of relativity.

But simultaneously, as we shift our awareness towards the universal network of intelligent plasma fields, this same human consciousness is enabled to travel outside of time and space, and experience mystical, non-dual realms beyond the boundaries of the rational mind. We exist in all things, both within time and outside time, within space and outside space.

Plasma cosmology explains how the entire universe seems to operate like a single pulsing organism, instantaneously transmitting information back and forth across vast distances irrespective of the speed of light. It explains how the physical universe itself is an emanation of cosmic consciousness, and how everything that can be perceived within time and space is shifting back and forth between form and formless realities.

This theory also explains the synchronicity of galactic and cosmic events across time and space. When a galactic superwave pulses outwards from the galactic center we feel it instantaneously. When the Sun experiences a flare we feel its effects on Earth immediately. Through the principle of plasma resonance, the 12,000-year heartbeat of the galaxy that we referred to earlier could signify a pacemaker for coordinating events, not just within our own galaxy or Solar System, but across the entire cosmos.

Douglas Vogt believes this is true, and that all perturbations and interactions within the cosmic web ultimately trace their origins beyond the web. He referred to this formless, or causal realm as the Diehold, the same field that the ancient Hindus referred to as Brahman, and others simply as the Source field, Universe, God, or Cosmic Consciousness. Matter and energy are not so much created out of this realm of Consciousness, but are constantly oscillating back and forth between states of form and formlessness! It is only Maya that causes the rational mind to believe that these two states of existence are separate from each other.

This theory explains how we as multi-dimensional entities can cross between timelines, and create multiple dimensions of reality, based on where we choose to place our focus. Just as Schroedinger's cat is neither dead nor alive until we open the box to take a look, so there is no fixed reality out there until we, in multi-dimensional awareness, choose to make it so.

On a practical level, an understanding of this theory allows us to reach beyond the finite resources of our dying Sun, and tap directly into the plasma currents that constantly surround our Earth and Solar System. Whether we utilize these currents to enhance our biological capabilities, tap into this infinite source of power to replace our dependence on fossil fuels, or allow these understandings to shape our philosophical, psychological, scientific and spiritual paradigms, it is crucial to our survival and continued evolution in these pivotal times.

Ultimately, this model also provides me with a blueprint for how homo luminous could actually emerge. Our nervous systems have evolved to a point where our biology has become extremely receptive to our vibrational consciousness, which in turn has much to do with our collective conditioning.

The Plasma Cosmology paradigm empowers us to seek our true identity as universal consciousness, the Self that permeates all things. Once we know ourselves as the Self, higher levels of the mind can begin to integrate with the rational mind, allowing us the experience of transforming and shaping our perceptions of reality in the world of physical matter as well.

The paradigm shift from a gravity model of the universe to an electric model of the universe is just as revolutionary as the shift from a geo-centric model of the universe to the Copernican model of planets spinning around a Sun, or from Newtonian physics to relativity and quantum physics. Every paradigm shift we make helps our evolution by learning to perceive the universe differently. For as we perceive, so we become.

CHAPTER 49
THE UNIFIED FIELD

Once we understand the basics of the electric universe theory it is only a tiny step further towards recognizing that the universe is conscious, and infinitely alive.

We have spoken earlier about the matrix of duality locked within the magnetic fields of the earth. A matrix is like a template, a very subtle network of filaments or fibers that holds a certain reality in place. This matrix of duality is a mental construction, and has nothing to do with physical reality. As most species on earth understand, it is possible to live in the physical world without being mired in duality.

The physical matrix does have to do with magnetic fields, however. If electrical fields have to do with the wave function of matter, magnetic fields are associated with its particle nature. As we saw in the last chapter, the science of plasma cosmology tells us that 99.999 % of matter exists in a pre-condensed state known as plasma, which consists of charged particles floating in the vast regions of space, held together through electric filaments which span across galaxies in a macrocosmic form very similar to neural networks within the human body.

What is it that causes plasma, floating freely in an electric universe, to coalesce into the states of matter familiar to us as gases, liquids and solids? Is it magnetic fields that condense plasma into visible states of matter, generating in turn the forces of gravity?

The physical universe is but a tiny subset of the vast ocean of Consciousness that exists beyond the original matrix of time and space known to the ancient philosophers of India as Maya, the cosmic lattice through which all Creation takes place. Consciousness is independent of Creation, but also permeates everything that exists within Creation. Creation did not happen in one single instant of space-time known as the Big Bang, but is a continuous movement of Consciousness between states of rest and motion, formlessness and form.

As Consciousness passes through the veil of Maya, electrical fields are created, which hold the entire physical universe together in a single cohesive web of light. All things are instantaneously linked together through this web, as demonstrated in the electric universe theory. Electrical discharges across vast distances leads to the emergence of plasma, also known to the ancient seers as ether. Plasma is the primal state of matter, but can pass freely across the veil of Maya. The universe is an open system, with plasma capable of entering creation and then dissolving back into the formless void trillions of times every second.

As plasma interacts with electrical fields, local magnetic fields are generated, which in turn gives rise to condensed elements of matter in the form of gases, liquids and solids. Consciousness and matter exist within the same continuum, each simply being an extension of the other. It is only the filters within the rational human mind that perceive them as apparently separate from each other. It is these same filters which eventually give rise to the collective perception of separation and duality, the source of all human ignorance and suffering.

Just as Consciousness has dual properties of form as well as formlessness, so it is also within Creation. We are familiar with the dual properties of waves and particles within photons of light. The same holds for the polarity between energy and matter. Electric fields are linked with waves or energy, whereas magnetic fields are linked with particles or matter. It is when these two fields are brought together that life is created within Consciousness, and Consciousness can incarnate within life.

This truth became very clear to me in my study of Inka shamanism. Inka cosmovision recognizes that the entire universe is alive with Consciousness, and that there is a single flow of living energy, or sami, permeating the universe. The Inka shamans further recognize that there is a reciprocity between all things, known as ayni.

The Inkas refer to three worlds of Creation. The one most familiar to us is the kay pacha, or the middle world, the narrow world of the five senses governed by the rational mind. Above this is the hanaq pacha, the infinite worlds of spirit, accessible through levels of the mind beyond the rational; and below this is the ukhu pacha, the earth realms woven together by a single web of life, and accessible to us through the deep wisdom of the body, also known as the cellular mind.

In my studies with Inka paqos, Don Juan Nunez del Prado and Don Ivan Nunez del Prado, I was taught to access the powerful electrical currents constantly pouring down from the upper world, and the magnetic currents constantly moving up from the lower world. As these two streams interact, a refined yet powerful matrix of light is created within the causal, subtle and physical bodies through which the Self can incarnate.

This was an extraordinary discovery for me. Much of my spiritual journey in the past had been about attempting to anchor divine light into the physical body. But I had always seen this as a long, slow, and difficult process. Once I became aware of this matrix of light, however, I experienced that the Self could very easily incarnate within these bodies of creation as long as I remained open to these electrical and magnetic flows of sami.

As above, so below. As Consciousness becomes capable of incarnating within the human form in the presence of subtle electrical and magnetic flows of sami, I realized that the electric universe theory provides an explanation for how the same Consciousness that exists beyond the physical universe also permeates the physical universe. Electromagnetic fields are templates through which Consciousness incarnates as matter.

As the advaita teachings indicate, Consciousness is non-dual. The physical world may appear to be separate from spiritual realities, just as waves of light may appear to be different from particles of matter. This separation exists only in the mind of the perceiver, however, and as we shift our perceptions we begin to experience what yogis and shamans across the ages have always known, that there is but a single, undivided reality pervading the entire universe and beyond.

This reality is the Self, the Unified Field of Consciousness that creates all things and moves through all things, within which we find our deepest essence and truest home, the Force that simultaneously drives Evolution as well as Involution, the Source of all qualities of beauty, wisdom, love, creativity and power, existing within as well as beyond the apparent veils of Creation.

In the understanding of this truth, a deep force began to weave itself through my bodies, bringing me to a point of absolute stillness within a great wave of cosmic expansion. The rigid boundaries of the personal ego dissolved. A viscous web of light permeated the three bodies, extending deep into the earth and wide across the sky. A presence far greater than the personal ego began to gaze outward through mind and senses. Was this an aspect of the supramental force descending into the heart of Mother Earth at this time, preparing us for new worlds to come?

CHAPTER 50
COSMIC DNA

As we begin to grasp the essential nature of the Self and of human evolution, we come to one of the most significant paradigm shifts of all, and this has to do with the nature of our biology itself.

We have long known that the growth and evolution of life has to do with genetic codes known as DNA. Humans share DNA codes with every form of life on Earth. Indeed, humans and apes share 99% of the same DNA, while humans and mice share 90%. The further apart we get the less DNA we share in common, but even so every species has more in common with each other than not, even if some have long gone extinct.

What exactly is this DNA? From the perspective of mechanistic biology, these are the building blocks of life, nucleic acids strung together in highly complex chains of information, which replicate according to fixed laws of biology. Just as atoms gravitate together to form molecules, so nucleic acids group together in specific patterns to create life.

How did life first originate? In the last chapter we looked at assumptions about how matter first originated, and how this assumption has now been shown to be faulty. We tend to perceive reality through filters of cause and effect, and therefore assume that there has to be a starting point to biology just as there had to be a starting point for time, space and creation.

But what if the replication of DNA has less to do with biochemical reactions than with informational fields, less to do with physical molecules than waveforms of pulsing micro-currents that are coded through the entire universe?

Lance Schuttler, researcher for TheMindUnleashed.com, tells us that we are surrounded by pulsating waves of invisible genetic information, which create microscopic gravitational forces that pull in atoms and molecules from their surrounding environment to construct DNA.[1]

Perhaps it is possible to go even further now, in light of the electric model of the universe, and say that these microscopic gravitational forces themselves draw energy and sustenance from the electrical plasma that fills the entire universe.

Recent experiments bear this out. Schuttler refers to Nobel prize-winning biologist Dr. Luc Montagnier, who placed two sealed and sterilized test tubes next to each other, each of which contained only pure water. One of these contained small amounts of DNA floating in the water. Both tubes were kept within a weak seven Hz electromagnetic field. Within seven hours the same DNA had begun to grow in the tube containing only sterilized water!

But that's not all. In his book, *The Source Field Investigations*, David Wilcock reports an experiment where a petri dish of sterilized water is placed in conditions where it could receive magnified starlight in a weak plasma field. As photon energies from the cosmos penetrated the petri dish, DNA started to form, just as reported by Dr Montagnier. How is it possible that new DNA was being created within this petri dish, where there was no physical proximity to the source of this parent DNA?[2]

Meanwhile German microbiologist Fritz-Albert Popp made another significant discovery, noting that DNA is a carrier and transmitter of light, and that each molecule of DNA was capable of receiving and storing up to 1,000 photons within itself. This phenomenon is known as bioluminescence, and the photons held within a DNA molecule are known as biophotons. The health of an organism has to do with how many biophotons it can carry within itself.[3]

Perhaps we are back in the world of plasma cosmology. If the entire cosmos is held together through micro-electrical plasma fields, this DNA is not travelling across distances, but is simply vibrating across a living web of life that weaves all creation into a single unified field. The bioluminescence exhibited within the cells of our body is simply the expression of these luminous strings of plasma that permeate the entire cosmos.

Implications of this paradigm are immense. It means that we are not isolated in the universe. Our Earth is not the only planet where life is capable of existing. Life evolves in great profusion within the gravitational forces of a single planetary organism. But DNA can also travel across vast distances of physical space.

This understanding of cosmic DNA suggests that the universe is constantly creating biological life, along with conditions where life can thrive, wherever and whenever it can. Recall Dr LaViolette's assertion that new creation originates at the centers of galaxies, and that these waveforms of creation travel across space in 12,000-year pulses known as galactic superwaves.

Whenever these pulses travel across the plasma medium of the cosmos, and wherever they find the basic conditions to thrive, micro-gravitational waves begin gathering atoms and molecules together around these packets of plasma-encoded information to create new DNA, and thus, new life.

As we noted earlier, this is not unlike the mechanism through which crop circles are being created in recent times. Even more interestingly, a study of plasma physics can perhaps help us understand better the mechanisms used by advanced space-faring civilizations to travel across vast stellar distances.

As we have seen, a new superwave of galactic information is preparing to pass through our solar system. Our Sun is going into a quiescent gestation period in order to receive this pulse of new creation. I suspect that as we come out of this gestation period sometime within the next 30 years, and into the final phase of a magnetic reversal, Gaia will be able utilize these new genetic frequencies to rebirth Herself in a profusion of extraordinary new life, including the next human species.

I suspect there is an akashic library of DNA held within the plasma field that includes every single species that has ever existed anywhere in the vast cosmos. It is up to Gaia, during times of magnetic reversal, to choose which ones of these myriad forms She wishes to invite into each cycle of evolution. We ourselves are part of Gaia too, and are therefore not separate from this process of new creation.

We cannot perceive the physical and the metaphysical apart from each other. Gravitational forces represent the particle aspect of creation, while electromagnetic forces represent the wave aspect of creation. We might refer to this as matter and spirit, but we hold both of these forces inseparably within ourselves. Once we recognize their co-existence we move beyond the illusion of Maya and free ourselves from the limitations of duality.

We are both creator and creation. This emerging new paradigm represents our map of enlightenment, our journey into homo luminous. It is in integrating these forces that we achieve the supramental descent, a unified consciousness that the sun-eyed children of a marvelous dawn have come to initiate here on Earth!

CHAPTER 51
RAINBOW BODY

How do we apply the principle of bioluminescence to our evolutionary journey? Our bodies are capable of holding light, light not just in the form of illumination but also information. This information, embedded everywhere within the luminous web of the plasma field, becomes available to each one of us through biophotons within our DNA.

We are simultaneously receivers and transmitters. As nodes in the great web of life, we touch the infinite, and carry the entire plan of cosmic evolution within our genes. We responsively touch back, taking this plan of evolution yet further.

In 1991, Michael Talbot, inspired by the work of theoretical physicists Karl Pribram and David Bohm, wrote a book titled *The Holographic Universe*. Highlighting parallels between ancient mysticism and quantum physics, it offers a model of reality suggesting that the universe functions like a hologram, and that telepathy, precognition and other paranormal phenomena are a direct consequence of this holographic model of reality.

A hologram is a three-dimensional image created by two interacting beams of light. The first beam encodes an image, hidden from physical sight. The carrier beam interacts with this first beam, revealing this image as a 3 dimensional hologram. What is interesting is that, unlike an ordinary photograph, each piece of the hologram contains the entire image.

249

When the Vedic seers talked about Indra's net, they were referring to the cosmos as a holographic reality, and recognized that each of us is a pearl reflected and being reflected by the entire universe. Once we recognize this, the veils of illusion are rent apart, and we become the guardians and gatekeepers for new genetic pathways.

The Inka shamans of Peru spoke about three worlds that we simultaneously exist in. The upper worlds reflect this sense of holographic unity. The lower worlds represent the cellular consciousness of matter responding to the cosmic forces of involution. And the middle world represents our individual perceptions, how we respond to this vast sea of information all around us. Most of the time, trapped in a narrow band of physical senses and the rational mind, we remain separate and ignorant of these mighty forces and infinite creative potentials.

We tend to see our physical bodies as separate from the various levels of subtler bodies that link us to the upper and lower worlds. But in the shamanic view, these worlds are not separate from each other, and we cannot therefore make a meaningful distinction between the physical body and non-physical light bodies.

From the perspective of plasma cosmology, *all* our bodies are light bodies, the physical body no less than the emotional, mental or causal body. Likewise, Gaia Herself exists simultaneously in all three worlds, keeping her in continual resonance with the Sun, stars and infinite galaxies in a great unfolding dance of life.

The Peruvian shamans, or paqos, referred to the Earth as Pachamama and to the Sky as Pachatata. But the Sun, the moon, the stars and our own material bodies are equally aspects of Pachamama, and the same universal spirit moves equally through all things as Pachatata. This again mirrors a holographic view of the universe, a perspective that allows us to engage directly with these forces of life in a reciprocal dance that the paqos referred to as ayni.

Do we truly grasp that we are holographic mirrors of Gaia? The whales taught me about this, as I felt within their consciousness field the ecology of an entire planet. This is no less true for any of us. We are subconsciously connected to the totality of Gaia, being impacted as well as impacting back in each moment of existence.

As we learn to hold more bioluminescence within our bodies, we begin to progress from unconscious entrapment towards a conscious awareness of these fields. We begin to sense layers of reality far beyond what is available through our ordinary senses, mediated through the rational mind.

The yogis of ancient India understood that our ultimate identity was the Self, and that this Self incarnated into human existence by clothing itself with five distinct interconnected bodies known as Mayakoshas, or apparent sheaths.

The Annamayakosha was the food body, our physical structure. The Pranamayakosha was the energy body, gathering life force from universal fields to feed the annamayakosha. Then was the Manamayakosha and Vignanamayakosha, comprising mind, intellect, emotions and the sense of personal self we refer to as the ego.

Finally, at the subtlest levels of expression, is the Aanandamayakosha, or the bliss body, whose function is to keep us connected to the infinite flows of life constantly emanating from the formless ocean of cosmic consciousness.

The illusion of Maya creates a veil, through which our human consciousness is blinded to its true identity as universal consciousness. This veil is manifested in human experience as a short-circuiting of plasma flowing between these five bodies, disconnecting the flow of Ayni, and leading to a chronic experience of separation and duality. This is the root of all human suffering.

The ancient yogis of India had devised internal practices for reconnecting with this flow of life, and with the higher realms of creation. Ultimately this resulted in building the mind of light, which could then create an entirely new body of light known as the nirmana kaya, or the rainbow body.

This is the radiant body of light, the immortal body, which can transcend all limitations of time and space. Although composed of the same five elements of matter, the vibrational frequency is different, and therefore responds to higher laws of the universe than what most people can understand or respond to.

Telepathy, teleportation and bilocation become normal for one who has achieved the nirmana kaya, as do phenomena such as shape shifting and time travel. At this stage it becomes possible to reanimate someone who has died, or to transfer one's own consciousness into another person or animal, whether dead or alive.

Most importantly, in breaking past the veil of Maya, one who has perfected the nirmana kaya is also directly linked with the body and consciousness of the Earth, and can participate directly and consciously in Gaia's evolution. Since they are no longer subject to cycles of forgetfulness and remembrance, they can therefore be completely focused in the great work of cosmic involution, where the consciousness of creator, Purusha, merges with the material form of creation, Prakriti; the end and ultimate purpose of our current cycle of yugas.

There are many adepts and yogis who have mastered the *nirmana kaya* over long ages of Earth evolution, as revealed in the Sun Men of South America, the ascended masters of old Egypt, the nath lineage of ancient India, the rainbow lamas of Tibet, and the Taoist immortals of China.

But our work now is different. It is no longer individuals within a species, but an entire species of humanity that is invited to experience the Nirmana kaya. And since humans represent the nervous system of Gaia, it also represents the possibility of Gaia Herself taking the rainbow body, sparking a new cycle of cosmic evolution.

This is the great work of Supramental Descent that Sri Aurobindo and the Mother have been pioneering, and that we are being called upon to continue.

CHAPTER 52
PANDORA'S BOX

R ichard Bach's book, *Illusions*, is one of my favorite books of
all time. It makes scattered references to a master's manual. I
would often wish there were such a manual — one that
would answer all the questions of life, sort of an antidote to
Pandora's Box, which had been so tragically emptied.

Eventually I realized there was no such book to be found, that
our searching for this book itself was enough, much like the
ancient alchemists who, in searching relentlessly for the
philosopher's stone, ultimately found it within themselves.

When I started writing this book I wanted to piece together
various elements of our ancient future. Some answers may be
found here for those who ask. But more important are the
questions we ask ourselves. In the questions lie the answers, if
we know how to ask them and if we live with them long enough
to let them shape our lives.

And what was this Pandora's box? There is an ancient myth
about a Greek god who in the long ago days before suffering was
ever invented, presented a beautiful maiden named Pandora
with a mysterious golden box. As in the story of the Garden of
Eden from a parallel ancient tradition, this god made a big
ceremony of presenting her with this wonderful gift, while at the
same time forbidding her to ever open the box.

So inevitably, of course, the time came when her curiosity overcame her caution and her fear. Just as her mythic sister Eve, had promised herself just a tiny bite of the fruit from the tree of knowledge of good and evil, Pandora promised herself just a tiny peek inside the box. She cautiously opened the golden lid. Too quick to be contained, every manner of confusion, fear, and human evil flew out and entered the world. Horrified, Pandora quickly shut the box tight, but it was too late. The evils had already been let loose.

Unfortunately, the last thing left in the box was Hope, the only thing that could have contained and healed our human suffering. Ever since that time, Hope has remained shut in the darkness while all the ills of our human condition flew unfettered through long centuries of painful drama. Dare we risk opening the box once more? Dare we find hope sufficient enough to look this god in the eye and, laughing, recognize our own face?

Stories always change when wrapped in the freshness of dew. We trap ourselves in little boxes and die, not even knowing we are trapped. If I wrap myself in stories of yesterday, this eternal moment vanishes in mist. Why imprison myself so?

Consciousness has no form when it is freed. Love does not need a story or a reason. I remember myself as creator and all creation becomes joy. Nothing else is real.

PART V

THE NEW EARTH

This evening the Divine Presence, concrete and material, was
there present amongst you. I had a form of living gold, bigger than
the universe, and I was facing a huge and massive golden door which
separated the world from the Divine. As I looked at the door, I knew
and willed, in a single movement of consciousness, that "the time has
come," and lifting with both hands a mighty golden hammer I struck
one blow, one single blow on the door, and the door was shattered to
pieces. Then the supramental Light and Force and Consciousness
rushed down upon Earth in an uninterrupted flow.

The Mother, February 29, 1956

255

CHAPTER 53
FUTURE LIFE
PROGRESSION

Earlier in this book I shared the discoveries of clinical hypnotherapist Chet Snow as he progressed people forward in time. In states of hypnotic trance, they seemed to experience one of three very distinct possible scenarios. As a transpersonal psychotherapist, I have had the privilege to explore these possibilities with clients. I share below some excerpts that appear to be relevant to our evolutionary journey today. This particular client, a young woman named Lila Hart, seemed to experience the beautiful green Earth scenario leading into the experience of directly embodying her light body.

So where are you now?

I am not sure if this is Earth after Shift or a different planet. I am glowing... same lifetime... so it must be Earth... everything is glowing. Things don't look in their material form anymore. Things seem less solid, less dense.

How did this happen? Go back to just before this Shift and fast forward.

I see darkness... lasting three days... three days in linear time. I can't even see the Sun. It's like there's a solar eclipse. My human body is in a deep slumber. I think many people are asleep. There is a darkness on Earth. But there are some other beings on the planet. They are all in their light body. They came while we were

256

in darkness, while we were lying in slumber, to assist us in the process of going through the Shift. They are with us. Earth is in a slumber as well. Earth is undergoing some changes as well, not just us humans.

What kind of changes?

The vibration of Earth is changing. Just like our bodies our transforming so is Earth... so beautiful... so beautiful.

(At this point, Lila began to experience extreme discomfort and difficulty with her breathing. It seemed that her physical body was not able to handle the intensity of this light. I brought her out of trance at this point. A few days later I guided her to a place where she could connect with her light body, and we went back into that same world.)

So you're in a body of light after the Shift. How does it feel?

Feels wonderful, I love this body... hmmm... it feels so nice, feels really good.

Does it feel very different from the physical body you had before?

Yes, my heart chakra feels so... I'm feeling so much more... I'm feeling light, I'm feeling the Spirit in all that is in matter... hmmm... it feels so nice... my body is this white-light-bluish glow.

It's still a body?

It's a light body, not solid matter like the flesh body, a different vibration. I can feel with my heart, feel things without touching them.

What kind of things?

I can feel the light in plants and trees, I can... everything looks very beautiful. Things are much more dynamic, not so static and solid as they are on the physical planes of matter.

And your senses can reach out further?

Yes. I see the energy flow like... if I was to look at a tree in the human body I would see outlines and leaves very sharply, but the energy that flows through it I could not see. That's how I see things now. I see what flows through, what's behind the form, behind the physical appearance, what lies at core, the true essence.

What about hearing?

It's different. It's a different kind of hearing, hard to describe it. More like hearing vibrations, let's say energy can vibrate on different levels, different frequencies. I can hear them now, that's how I hear...

Okay, continue on. What other senses do you experience?

The feeling thing is so strong. My main sense is feeling. That's how I tend to experience things, just feeling them, feeling this source, feeling Spirit and all that exists, feeling the inter-connectedness between all things, communicating with this inner connectedness.

And touch is very overwhelming, a very profound and deep intense experience. It's nothing like a human touch. When I am in my light body and if I were to touch another being, another light body, like holding hands or touching, it's very intense, it's because everything exists on this feeling level... it's like blending together your energies... its really profound.

So how many beings are there in this world that you see?

I cannot tell. Everything feels one... like there is only one being. We may exist separately in these forms of light but we're not separate, we are one being.

So describe your body. Is there any gender?

No.

How does your body look?

It still resembles the human body in some ways, like in contour. Much taller. There's differences in light bodies. Some are very, very tall and slender, hmm, yes, others seem quite short.

So what's the difference based on?

There are different beings from different star systems. They are not bound here. They can come and go.

So let's go back to the Shift itself. You mentioned earlier there were 3 days of darkness and people were going into different dimensions. So what actually happens during this time?

If they are vibrating on a particular dimensional frequency they shall go... it's more like being called to... attracted to a certain reality based on your vibration... being called there... you long to go there.

So which dimensional frequency are you experiencing now, if you could name it, speaking in very relative terms?

Hmmm... this plane of existence... vibrates on the seventh-dimensional frequency.

From this dimension do you have access to other dimensions that are vibrating at lower frequencies? Can you visit and return?

I can see them. The only dimension I am allowed to go to... I am only allowed to interfere with third-dimensional affairs. The others I have to let be.

(As she focused on the third-dimensional world, she began to see some rather catastrophic events happening: volcanoes, ash clouds, landmasses breaking and shifting. In the midst of all this she sees a light).

There is a light on Earth. I don't know what it's doing there. I was under the impression all the lights had left by now. It's a being in the light body. I'm being called towards this being.

So who is this being?

It's strange. I am not sure what... I am seeing simultaneously Earth in chaos and then this green paradise which is Earth, unravaged, untouched... they are existing simultaneously.

Different dimensions?

No.

Two parallel realities?

Yes. I am trying to figure out why these two different realities are co-existing.

Go to this being that you see. Ask him to explain.

259

They are co-existing because this is before the Shift. It's during these three days that it is decided what will happen. They are still there simultaneously because either could be the third-dimensional Earth. This being is here to ensure that the green untouched Earth will be the reality which will prevail over the other chaotic disruptive one.

So everyone gets to experience this green beautiful Earth?

No.

Okay, so what happens?

I'm feeling there are several possibilities. Those souls who have not evolved enough to go through the Shift can continue life on different planets as well. They won't be on Earth.

And Earth has two choices — to be destroyed and extinct or to renew herself?

Earth will not become extinct. Even if the chaotic and destructive Earth would prevail it would not go extinct, it would not die. The planet would revive after all the rupture and chaos. The Earth survives no matter what.

So it would be like a new incarnation of Earth?

Yes. The timelines and possibilities for what is happening after the Shift and the days before the Shift are not definite. As human consciousness evolves and grows up to that point the scenarios keep changing.

So the longer we wait before the actual Shift the more likely it is that human consciousness would evolve further, so that we will choose the green Earth?

Yes.

So when in human time do you see the most likelihood of this Shift happening?

It may take centuries before humanity has evolved enough. However, the Shift will take place sooner, much sooner. Help will be given. No dates.

Has this happened before — these times of Shift?

Yes.

What's the difference between this time and previous times?

A large mass of people will go through this Shift. It is designed so that many, many souls can experience this accelerated growth. Many, many beings are involved and working for this. It is like a project, like an experiment to see what will happen, how this Shift will progress with so many humans, so many souls incarnated in human bodies.

Is this an experiment that involves other galaxies and universes?

Yes.

So in a sense when the Earth experiences the Shift to the green planet it changes the vibrational frequency of the whole material universe?

What do you mean by the whole material universe?

The galaxies, stars, visible and invisible.

If Earth's vibratory level is successfully changed, even this third-dimensional Earth will go up in frequency and vibrate at a higher level. Should this succeed, it will allow access to different star systems.

So what's the purpose of this evolution right now — experiencing ourselves incarnated in solid matter, forgetting the connection with Spirit, feeling separate in time and space, feeling limited, feeling the pain and suffering of duality?

It is only to realize that none of it is real. There is no suffering in consciousness. There is no separation in consciousness. This forgetfulness allows for experiences. If we were fully conscious coming into physical matter we would not experience suffering, even if suffering is only illusory. Yes... once we are able to gather these experiences the joy is indescribable... even though in our limited states we may see them as suffering. These journeys, these experiences are very valuable to soul, to Spirit, yet we do not understand it when embodying these human vehicles... we do not remember with what joy we took birth in these bodies.

* * *

Perhaps this is a good place to end for now... reconnecting with our purpose... reconnecting with the joy that permeates our journeys on Earth!

Next, I would like to share a profoundly hopeful vision experienced by Grace Sears, who has travelled through India many times in the past few years.

CHAPTER 54
A VISION OF HOPE

By Grace Sears

India has been a place of great activation for me. It is a place where I experience spiritual clarity and connection to other realms and other times. I feel a knowing that all things exist simultaneously, and that we have access to what seems other-dimensional if the veils of 'reality' have sufficiently thinned. I feel the Great Mother often in India. I hear her voice and sense her movements.

I would like to share an inner vision that I experienced here some time ago. It is one of the profound gifts I have received from Mother India. This vision has been of deep comfort to me. I am sure it will serve to comfort others during these tumultuous times of transition and change.

It began in south India in January of 2009 with a powerful mystical experience. I found myself observing my body from a higher perspective. My identity existed in a state of unity that was within and yet beyond the physical body. I contemplated the symmetry of my body — two legs, two arms, two eyes, two lips... Was this the reflection of the duality intrinsic in human form? I touched my face in wonder and amazement.

I proceeded to weave myself back together in this state of higher consciousness. I was in a cocoon of pale golden white light, dense but fluid, looking much like images of Osiris on the walls of Egyptian temples. My arms were bound, but my feet seemed to be visible within the cocoon, and my face was uncovered with a band of brilliant gold around it, radiating immense light and joy. I could not think in linear fashion, nor hold any thoughts, but I felt an awareness that I was done. There was nothing left to do on Earth, nothing to have, become or experience. There was nothing that interested me, nothing I desired on a human level anymore.

For several days, I remained in this state of completion, content to be without desire or direction. I was unsure whether this meant I had finished the tasks of this lifetime, or whether it was a transition to a different mode of existence.

A few days later, my friend Audemarie and I were in a taxi heading to visit Kiara. He was set to emerge from his kayakalpa retreat — a month-long rejuvenation program that involved cleansing with ayurvedic herbs while in darkness and solitude. Audemarie told me she was planning to go through the same treatment a couple months later.

As we sat together, I knew with a deep inner certainty that this was what I was being called to do also — to rejuvenate and purify my body to hold more of the light of the new times to come.

When we arrived at the ayurvedic center, I spoke with the doctor and we agreed that I would go into the kutir — the womb-like hut designed specifically for kayakalpa retreats — for one full cycle of the moon.

In May of 2009 I underwent my kayakalpa retreat. It was a powerful experience. I had many visions and awarenesses, amplified by twenty days of intense kundalini activity. Often, as the kundalini rose to my head centers, I would experience inner visions of great clarity and knowing. Sometimes I experienced differing visions on each side of my head.

I oftentimes 'visited' a Buddhist monastery high in the mountains, perhaps somewhere in Tibet or Bhutan. I could see individual monks in perfect detail. I saw the way the light touched their faces and reflected in their eyes. I heard the music, the voices, and even the wind.

A Vision of Hope

Often in these visions I would see an immense golden statue of Buddha. Worshippers in beautiful silks were seated in a semicircle in front of the statue with their arms and foreheads on the floor.

On the second to last day inside the kutir, I saw a Tibetan-looking man wearing a hat with earflaps struggling up a hill in a heavy snowstorm. He seemed to be pulling something with great difficulty. This vision was on the left side.

On the right side, I saw the huge golden Buddha and the silent worshippers in their beautiful silks. Chinese soldiers were running toward them dressed in black with strange hats and weapons. They seemed invincible in their hatred of the Tibetan monks. They were filled with an intense and destructive rage. They were so convinced that they were right and unable to accept the truth of the other. I was aware that nothing could be done from this plane of consciousness to stop their advance.

On the last day of my retreat, as the kundalini rose once more, I saw the same visions again — the golden Buddha and the Tibetan man struggling in the snowstorm. Then, as I watched, a second Sun unexpectedly rose in the sky, but from the west. Instantly, the blizzard stopped and the snow melted. The mountainside turned fresh and green, dappled with wildflowers. The man walked with ease up the path leading his donkey, smiling. I could see he was on his way to the peaceful monastery on the crest of the next mountain.

As the rays of the second Sun struck the Chinese soldiers, they were driven rapidly backward — shriveling and disintegrating. Then, as I continued to watch, they formed a second semicircle, much larger, around the worshippers, also dressed in lovely colored silks, sitting in lotus position, silent and serene.

This vision gave me great comfort. I knew absolutely that we are taken care of as we move into a new time, a new world, which will manifest at the right moment in a seemingly miraculous way. What we are unable to change on this plane despite all our struggles and intentions is accomplished effortlessly, or rather, it simply IS, as human consciousness makes the Great Shift.

May we be blessed, grateful, and filled with joy!

CHAPTER 55
A NEW SUN

What is this second Sun that Grace saw in her vision? When the Mayas talk about what will happen just before we enter the new world, they say there will be some days of darkness followed by a new sunrise. As Aluna Joy Yaxk'in shares:

There is a new Sun coming up over an ocean, and there is nothing else... Just the ocean and the Sun... The Maya say they are not really sure if it is a brand-new Sun or just the energy of a new Sun, but they say a new Sun will rise. A new Sun is being manifested in spirit right now. The Ancestors are saying that this new Sun is being birthed into the world... It will take a while for this to manifest physically. When it does, we will be receiving energy in a direct way...

As with prophecies of the three days of darkness, prophecies of a second Sun are part of many other ancient traditions. How do we interpret this prediction in a way that makes sense to us in our modern age of science and reason? There are many rumors and predictions these days about super-dense planets, brown dwarf Suns, and, indeed, entire miniature solar systems blazing their way through our Solar System, wreaking havoc in their path. Is this the vision that the Mayas are referring to?

I believe that prophecy is often based on ancient memories, which wisdom keepers of many traditions consult to predict future events based on cycles in the past. If so, what does this double prophecy of three days of darkness followed by the rising of a second Sun mean?

My sense is that although the Mayas may be referring to a literal event, we cannot interpret this prophecy from a third-dimensional perspective of reality. Perhaps in order to understand what this prediction means, we need first to take a journey into the deepest layers of our subconscious mind, into the matrix we have referred to as the veil of illusion.

Eastern spiritual traditions have much to say about this veil of Maya. A simplistic interpretation of this belief is to say that the world out there is an illusion, and that the way out of this illusion is to renounce the world. Although this response may have worked for us in simpler times, we must understand that now the veils of illusion exist within our own consciousness. The world out there is not the problem — our interpretation of this world is, based as it is on subconscious structures of programming that keep us trapped in perceptions of separation and duality.

This veil of illusion lies deep within our subconscious awareness, and is literally programmed into the matrix of our cellular DNA. The frequency of this programming keeps us trapped in a third-dimensional perception of reality, which we consider the one and only reality of existence. The belief that third-dimensional reality is the only reality is itself the illusion and we cannot make it through the Great Shift as long as this programming remains within our cells.

In context of our current civilization, the veil of illusion refers to our perception of the universe through mechanistic assumptions alone. What if the deeper implications of string theory and plasma cosmology begin to sink into our collective awareness, shape a new biological species, reprogram sub-conscious beliefs about who we are and what is possible, overhaul our socio-economic and political systems, and even transform our sciences and technologies so we can meet the challenges of these times?

Something very interesting begins to happen during times of geomagnetic reversal. As I said earlier, as the geomagnetic field of the Earth collapses, the collective matrix of personal identity begins to disintegrate. Our emotional and mental bodies are no longer tightly bound to old patterns of subconscious conditioning.

We begin to unravel the dense veils that keep us trapped in third-dimensional density. New potentials of genetic information become activated, which raises the frequency of the energy matrix vibrating the cells of our body, in turn helping us perceive realities beyond the narrow limits of our known physical senses. Our consciousness begins to merge with worlds and dimensions beyond third-dimensional matter. We open to subtler levels of the mind. We begin to anchor the multi-dimensional frequencies of the soul.

All this change happens very quickly during this cocooning period, as our sensory perceptions go into hibernation during these days of darkness. Creative evolution is a quantum process that utilizes the enormous wave of incoming cosmic energy to shape new creation. In the absence of the subconscious conditioning that has kept us trapped within the veils of illusion for eons, we make a quantum jump of evolution into a new state of biological consciousness capable of integrating the multi-dimensional frequencies of our soul.

Perhaps we can now interpret this prophecy of the Second Sun in a somewhat different way. Most Suns exist as part of a binary star system. Our experience in this Solar System is unique in that we only perceive a single Sun. But perhaps our twin Sun has always existed on a higher octave of creation and we simply have not been able to see it yet.

Perhaps it is not a second Sun entering our Solar System during this time of the Great Shift but rather our twin Sun simply becoming visible to us in this dimension, as our senses expand beyond the narrow limits imposed by linear consciousness!

Or perhaps, as Walter Cruttinden explores, we already know this binary twin as Sirius, and just haven't wrapped our minds around this possibility as yet! Our journey back towards our twin star then is a reflection of our emergence into the next Satya Yuga, the age of light.

This new Sun is already manifested in spirit, say the Mayas. Whatever the Second Sun might mean to us, as we open to this radiance we will begin receiving and assimilating its energy in a direct way. Our bodies will be nourished directly from the energy of the new Sun.

We will learn to use the energy of this Sun to build internal technologies that directly resonate with our higher selves. We will learn to heal our bodies, create new worlds. We will travel freely through multiple dimensions of the universe. We will nourish our bodies directly from the fields of prana all around us. Outdated social, political, economic, and religious systems will fall away because they will no longer be needed.

As the veil of illusion dissolves, so too, does our subconscious matrix of fear, separation, and duality. We will experience this planet as one single interconnected consciousness unifying all things within the web of life. We will discover what it means to experience Creator-Consciousness within infinite dimensions of Creation!

The Mayas remind us that the new Sun manifesting on the outside is a reflection of a new Sun being birthed within. This is the soul consciousness being birthed as we pass into the doorway of the new Earth! It is the birth of a divine human, serving as a steward on this beautiful blue and green planet, inspired by a constantly flowing well of creative impulse. Rather than reacting instinctively in predetermined subconscious patterns, we will respond in each moment with the full intensity of our multidimensional presence, nurturing and honoring each strand in the web of planetary life.

Whatever is real from the old Earth will remain. Experiences that feed and enrich our souls, memories of beauty and love, all of these will remain. Nothing is ever lost in evolution. Old structures are simply included and transcend to a higher octave of life. We will experience life in the same dimensions of matter, simply vibrating to a higher frequency of light. We will engage the world around us through similar personality structures, but harmonize directly to the consciousness of our souls.

This is the goal of creative evolution as we transition to the New Earth.

CHAPTER 56
GAIA LUMINOUS

The only thing that moves me these days is the overwhelming passion I feel for Gaia's evolution. Our current human species has forfeited its place in the web of life, for we have worked hard to destroy the very foundation of our existence, this Earth we call our home.

And yet Gaia has birthed us for a reason. We are part of an experiment to weave creator consciousness deep into the heart of matter. Perhaps the experiment has not worked as well as it might, but it is not over yet. We are a species in transition, and a greater dawn is at hand.

In May 2016, while on retreat in the magnificent Black Sea region of Turkey, I took some time to go into a shamanic space, communing with the heart of the Earth and the power of the Sky. I felt myself entering the primordial space before the existence of time, creator dreaming a new dream, Gaia emerging from the dream in a beautiful wild dance of creation.

Volcanoes erupted in the power of this dance as supernovas blasted across the skies and galaxies were spewed and swallowed alive through the magnificent forces of Maya. The wildness passed as Gaia matured, giving birth to a complex web of life in the passing of the ages.

Eventually came humanity, whose purpose of existence in the web of life was to serve as Gaia's link with infinity. But oh such a fragile species we were, unable to tolerate climatic extremes or the wildness of change. I felt how constrained She felt in her own energy, yet waiting ever patiently for an evolutionary force to move through time that even she herself could not fully imagine.

And then from far out in the supramental realms I felt something enter her womb, a galactic seed that arced across the electric fires of Shiva's dance. I felt a mighty golden hammer break open the door of humanity's illusion, calling us to enter the deep heart of creation with Gaia, and find our way home.

I then saw the sun-eyed children of a marvelous dawn, birthed from beyond the veils of perceived separation, bodies made beautiful by spirit's fire, shining with the power of love's desire for itself.

This was the new species awaited so long, being birthed on the other side of the magnetic reversal. It was human in form but not in expression, far removed from the anchor of separation and duality that has defined our matrix for so many eons. Its new matrix was the direct awareness of oneness that pervaded the infinite multiplicity of forms, so that the entire web of life could find its center, and radiate out even further.

This was not a human species alone, but carried the potentials of all species of life to evolve further. This was indeed the birth of a New Earth consciousness: not just homo luminous, but Gaia luminous, a planetary web of living light. I became Gaia then, and felt the mighty waves of delight that rippled through her being, as the dance of Shiva was once again renewed in a brand new creation.

I saw in the form of infinite consciousness that seed being planted. The supramental force was entering the structures of human DNA. We were entering the cocoon of this new creation. The human butterfly, the great cosmic child of a New Sun, was but a heartbeat away from manifesting fully in all forms and structures of creation.

I saw that the old world with its systems of soulless greed, mindless fear, and heartless oppression will very quickly fall away. There is no vital force sustaining these structures anymore. And the seeds of the New Earth are germinating just as quickly within the hearts of all who choose to awaken now.

The dreamer awakens from his dream, and all creation explodes with joy.

I am Shiva. I am. I dance with you my love, my beautiful cosmic Gaia. Let the lava of creative energy flow once more. Let the shackles go. I am here in the deep power of endless love. We dance the dance of creation; we create new worlds and entirely new species spring to life.

Homo luminous…Gaia luminous, you are the crown jewel in all the infinite waves of creation… for in you I see my face clearly, in you I express my love freely, in you the supramental force reveals itself fully… in you all separation dissolves into a single pulse of infinity…

A divine force shall flow through tissue and cell
And take the charge of breath and speech and act
And all thoughts shall be a glow of Suns
And every feeling a celestial thrill.
A sudden bliss shall run through every limb
And Nature with a mightier Presence fill.
Thus shall the Earth open to divinity
And common natures feel the wide uplift.
Illumine common acts with the Spirit's ray
And meet the deity in common things.
Nature shall live to manifest secret God
The Spirit shall take up human play
And this Earthly life become the life divine.

Sri Aurobindo (Savitri)

END NOTES

Preface

1: To watch a short documentary summarizing the basic themes of this book, please visit https://youtu.be/DeIThuIrBzs for an interview filmed in Egypt between Turkish filmmaker Ahmet Yazman and the author.

Chapter 13

1: Mark Lynas' fascinating book, Six Degrees: Our Future on a Hotter Planet (London: Harper Perennial, 2008), providemate begins to warm up, degree by degree.s a sobering analysis on what is likely to happen as Earth's climate begins to warm up, degree by degree.

2: International Climate and Energy Conference in Munich, 2011
3: Global Warming Hoax, Best Documentary Ever,
https://youtu.be/DJBDI7jVMqM

4: See YouTube video, Guy McPherson – Human Extinction Within Ten Years, https://youtu.be/zqIt93dDG1M, and Gary Null and Guy McPherson: Collapse of Civilization within next 10 years due to Climate Change, https://youtu.be/t580j47Y0Do

Chapter 14

1: See George Monbiot's website Monbiot.com for details.

Chapter 15

1: See Robert Felix's "Not by Fire but by Ice" referring to John and Katherine Imbrie's CLIMAP data.
2: See YouTube videos, Mini Ice Age has Begun,
https://youtu.be/9oUaWI2MQDY, Is an Ice Age Coming?
https://youtu.be/ztninkgZ0ws, Ice Age Shift,
https://youtu.be/-W6Lftgq8mg, Little Ice Age Big Chill Documentary,
https://youtu.be/JwuO4cXghBo, and Orbits and Ice Age: The History of Cimate - Dan Britt, https://youtu.be/xgNxF2HlN3w
3: Are we heading into another Ice Age, Peter Shinn interviews John Casey, https://youtu.be/lJBMk9iRDXE. Also, John Casey on Catastrophic Earthquakes Striking USA,
https://youtu.be/ZTJCY6M-3fY

4: See Robert Felix's website Iceagenow.com for details.

Chapter 16

1 See www.Ourhollowearth.com

2: The Earth is Growing, https://youtu.be/oJfBSc6e7QQ

Chapter 17

1: Dr. Alexey Dmitriev, "Planetophysical State of the Earth and Life," Tmgnow.com/repository/global/planetophysical.html

Chapter 18

1: See Paul laViolette's web site www.etheric.com for more details

2: Graham Hancock and Joe Rogan Discuss Randall Carlson's Paradigm Changing Research, https://youtu.be/sJA1D9tRs9E).

Chapter 20

1: Drunvalo Melchizedek, "The Maya Physical Poleshift Prophecy," YouTube.com/watch?v=8vWBxi2xTVc

Chapter 21

1: Sri Yukteswar, The Holy Science (Self-Realization Fellowship, Los Angeles, 1990), p. 8-10.

2: Walter Cruttenden, Lost Star of Myth and Time, https://youtu.be/1azoIQB2PMc

3: The Great Year Ancient Civilization Our Binary Solar System, https://youtu.be/XAP-Q-2HRpw

Chapter 22

1: Online excerpts from Douglas Vogt's book, God's Day of Judgment: The Real Cause of Global Warming, Vectorpub.com/Ch_08_GDJ.html

2: Randall Carlson, Earth Changes and the Precession of Equinoxes, https://youtu.be/B1hOUGu2wXw, and Cycles of Catastrophe and Cosmic Patterns, https://youtu.be/538tIdNbJJ0, and Catastrophism: A New History for Planet Earth, https://youtu.be/nhP_tI-AyIY

3: Colin Wilson, Atlantis and the Kingdom of the Neanderthals (Vermont: Bear & Company, 2006).

4:http://news.berkeley.edu/2014/10/14/earths-magnetic-field-could-flip-with in-a-human-lifetime/

5: Ben Davidson, Magneticreversal.org. Watch especially the videos, 5 truths about Magnetic Reversal, https://youtu.be/sIayxqk0Ees, What a Magnetic Reversal Means for Earth, https://youtu.be/pPqYfAHRPtk, and The #1 Risk to Earth, https://youtu.be/VVgUZv9ccyQ

6: Ben Davidson, Magneticreversal.org. Watch especially the videos, 5 truths about Magnetic Reversal, https://youtube/sIayxqk0Ees, What a Magnetic Reversal Means for Earth, https://youtu.be/pPqYfAHRPtk, and The #1 Risk to Earth, https://youtu.be/VVgUZv9ccyQ
7: The Gulf Stream Explained, https://youtu.be/UuGrBhK2c7U?list=PLyADScPoHm793IPZ5v9BMGCbQRTxAKa3A

Chapter 27

1: Georges van Vrekhem, Beyond Man: The Life and Work of Sri Aurobindo and the Mother (New Delhi: Rupa & Co, 2007), 75.

2: "Sri Aurobindo," Wikipedia, The Free Encyclopedia, En.wikipedia.org/wiki/Sri_Aurobindo

3: From Sri Gopal Bhattacharya's April 27, 1991 University of Cambridge lecture, "A Fragrant Flower of Cambridge: Sri Aurobindo: His Contribution to Humanity," Booksite.ru/fulltext/Aurobindo/bhattach/gopal1e.htm

Chapter 30

1: See flammedalterite.wordpress.com.

Chapter 31

1: Douglas Vogt and Gary Sultan. Reality Revealed: The Theory of Multidimensional Consciousness (Bellevue: Vector, 1978), 450.

2: http://www.collective-evolution.com/2017/07/22/synchronization-of-autonomic-nervous-system-rhythms-with-geomagnetic-activity-found-in-human-subjects/

3: Richard Bartlett, Matrix Energetics: The Science and Art of Transformation (Atria Books, 2009)

Chapter 33

1: Marianne Williamson. A Return to Love: Reflections on the Principles of A Course in Miracles (New York: HarperCollins, 1992), 162.

Chapter 36

1: Georges van Vrekhem. Beyond Man: The Life and Work of Sri Aurobindo and the Mother (New Delhi: Rupa & Co, 2007), 431.

Chapter 37

1: Please check out Barry Martin's website Luminousself.com

Chapter 39

1: For more details on free energy technologies you may wish to investigate websites such as Freeenergytruth.blogspot.com, Peswiki.com.Newenergymovement.org, Globalscalingtheory.com

Chapter 40

1: This refers to the age of the universe from the Big Bang perspective of the Standard Model of physics. Later in the book I refer to the Electric Universe model, which presents a different, and perhaps more accurate, version of Creation.

Chapter 43

1: Why the New Chronology Came into Being, https://youtu.be/4BfsvPRLcP8

2: Undeniable Proof Hitler wanted Peace, https://youtu.be/xGm_KeyWWYQ

3: See Gods of Money: Wall Street and the Death of the American Century by William Engdahl

Chapter 44

1: Along with John Perkins, please also examine the writings and YouTube presentations of John Pilger, Jon Rappoport, Paul Craig Roberts, Ron Paul, and Noam Chomsky.

2: Investigate this rapidly growing social movement at Thezeitgeistmovement.com and Thevenusproject.com

Chapter 48

1: Some of the pioneers of the Electric Universe theory include Hannes Alfven, Anthony Peratt, David Talbott, Donald Scott, and Wallace Thornhill. Please check the YouTube channel, Thunderbolts Project, as well as the work of Rolf Witzsche (Ice-age-ahead-iaa.ca) for more details. A comprehensive YouTube presentation is Thunderbolts of the Gods, https://youtu.be/5AUA7XS0TvA.

Chapter 50

End notes

1: Lance Schuttler, 'Scientific Experiments Show That DNA Begins as a Quantum Wave and Not as a Molecule, Feb 4, 2017. Omnithought.org/scientific-experiments-show-dna-begins-as-quantum-wave-not-as-molecule/5279

2: David Wilcock, The Source Field Investigations, and The Hidden Key: The Hidden Intelligence Guiding the Universe and You. Dutton, 2012

3: Yvonne Sangen, Biophotons: Source of Energy and Life Light. Utrecht: Ankh-Hermes, 2010.

BIBLIOGRAPHY

Abbot, John et al. *Climate Change: The Facts*. Stockade Books, 2015.

Allan, D.S., and J.B. Delair. *When the Earth Nearly Died: Compelling Evidence of a World Cataclysm 11,500 Years Ago*. Wellow, England: Gateway, 1995.

Alexander, Thea. *2150 A.D.* New York: Warner Books, 1976.

Anderson, Karen and Snyder, Barry Martin. *The Luminous Self Trilogy (Soul Awakening, Agents of Grace, and We are the Awakening Christ)*. Luminous Self Media, 2011-2012.

Argüelles, José. *The Mayan Factor: Path Beyond Technology*. Rochester: Bear & Company, 1987.

Bach, Richard. *Running from Safety: An Adventure of the Spirit*. New York: Delta, 1995.

Balsekar, Ramesh. *Peace and Harmony in Daily Living*. Mumbai: Yogi Impressions, 2003.

Batlett, Richard. *Matrix Energetics: The Science and Art of Transformation*. Atria Books, 2009.

Ibid. *The Physics of Miracles: Tapping into the Field of Consciousness Potential*. New York: Atria Paperback, 2009.

Bernard, Raymond. *The Hollow Earth: The Greatest Geological Discovery in History*. Secaucus: University Books, 1969.

Braden, Gregg. *Awakening to Zero Point: The Collective Initiation*. Questa: Sacred Spaces Ancient Wisdom, 1997.

Ibid. *Fractal Time: The Secret of 2012 and a New World Age*. Carlsbad: Hay House, 2009.

Ibid. *Choice Point 2012: Our Date with the Window of Emergence* in *The Mystery of 2012: Predictions, Prophecies, and Possibilities*, compilation by Sounds True. Boulder: Sounds True, 2007.

Broers, Dieter. *Solar Revolution: Why Mankind is on the Cusp of an Evolutionary Leap*. Berkeley, Evolver Editions, 2012.

Brown Jr., Tom. *Awakening Spirits*. New York: Berkley Books, 1994.

Caddy, Eileen. *The Spirit of Findhorn*. New York: Harper & Row, 1977.

Cannon, Dolores. *The Convoluted Universe*. Huntsville: Ozark Mountain Publishing, 2001.

Capra, Fritjof. *The Tao of Physics: An Exploration of the Parallels Between Modern Physics and Ancient Mysticism*. Berkeley: Shambhala, 1975.

Ibid. *The Turning Point: Science, Society and the Rising Culture*. New York: Bantam, 1982.

Carey, Ken. *Return of the Bird Tribes*. San Francisco: HarperCollins, 1991.

Ibid. *The Starseed Transmissions: An Extraterrestrial Report*. San Francisco: HarperCollins, 1991.

Ibid. *Vision*. San Francisco: HarperCollins, 1991.

Carroll, Lee, and Jan Tober. *The Indigo Children: The New Kids Have Arrived*. Carlsbad: Hay House, 1999.

Casey, John L. *Cold Sun*. Trafford Publishing, 2011

Ibid. *Dark Winter*. Humanix Books, 2014

Ibid. *Upheaval*. Orlando: Veritence Corporation, 2016.

Castaneda, Carlos. *The Teachings of Don Juan: A Yaqui Way of Knowledge*. New York: Pocket Books, 1968.

Cooper, Primrose. *The Healing Power of Light*. York Beach: Weiser Books, 2001.

Cruttenden, Walter. *The Lost Star of Myth and Time*. New York: St. Lynn's Press, 2005.

Fagan, Brian. *The Little Ice Age: How Climate Made History 1300-1850*. Basic Books, 2000.

Felix, Robert. *Magnetic Reversals and Evolutionary Leaps: The True Origin of Species*. Bellevue: Sugarhouse Publishing, 2008.

Ibid. *Not by Fire but by Ice: Discover What Killed the Dinosaurs ... and Why It Could Soon Kill Us*. Bellevue: Sugarhouse Publishing, 1997.

Findlay, Tom. *A Beginner's View of Our Electric Universe.* Grosvenor House Publishing, 2013.

Flem-Ath, Rand and Rose. *When the Sky Fell: In Search of Atlantis.* Macmillan, 1997.

Flem-Ath, Rand, and Colin Wilson. *The Atlantis Blueprint: Unlocking the Ancient Mysteries of a Long-Lost Civilization.* New York: Delta, 2002.

Forti, Kathy. *Fractals of God: A Psychologist's Near Death Experience and Journeys into the Mystical.* Rinnovo Press, 2014.

Fresco, Jacque, and Roxanne Meadows. *The Best That Money Can't Buy: Beyond Politics, Poverty, and War.* The Venus Project, 2002.

Gaspar, William. *The Celestial Clock.* Clovis: *Adam and Eva* Publishing, 2000.

Gilbert, Adrian. *Signs in the Sky: The Astrological and Archaeological Evidence for the Birth of a New Age.* New York: Three Rivers Press, 2000.

Greene, Brian. *The Elegant Universe: Superstrings, Hidden Dimensions, and the Quest for the Ultimate Theory.* New York: Vintage Books, 2000.

Ibid. *The Fabric of the Cosmos.* London: Penguin Books, 2004.

Ibid. *The Hidden Reality: Parallel Universes and the Deep Laws of the Cosmos.* New York: Alfred Knopf, 2011

Greer, Steven. *Hidden Truth: Forbidden Knowledge.* Crozet: Crossing Point, 2006.

Griffiths, Bede. *The Marriage of East and West.* Springfield: Templegate Publishers, 1982.

Ibid. *A New Vision of Reality.* New Delhi: Indus, 1992.

Hamaker, John D., and Don Weaver. *Survival of Civilization.* Lansing: Hamaker-Weaver Publishers, 1982.

Hancock, Graham. *Fingerprints of the Gods.* New York: Three Rivers Press, 1995.

Ibid. *Magicians of the Gods.* Thomas Dunne, 2017.

Hansen, James. *Storms of My Grandchildren: The Truth About the Coming Climate Catastrophe and Our Last Chance to Save Humanity.* New York: Bloomsbury, 2009.

Hapgood, Charles. *The Path of the Pole.* Kempton: Adventures Unlimited, 1999.

Hartmann, Thom. *The Last Hours of Ancient Sunlight: Waking Up To Personal and Global Transformation.* New York: Three Rivers Press, 2000.

Ibid. *We The People: A Call to Take Back America.* Portland: CoreWay, 2004.

Hawken, Paul. *The Magic of Findhorn.* New York: Bantam, 1980.

Hawkins, David. *Power vs. Force: The Hidden Determinants of Human Behavior.* Sedona: Veritas, 1995.

Houston, Jean. *Jump Time: Shaping Your Future in a World of Radical Change.* Boulder: Sentient, 2004.

Hubbard, Barbara Marx. *Conscious Evolution: Awakening the Power of Our Social Potential.* Novato: New World Library, 1998.

Hurtak, J.J. *The Keys of Enoch.* Santa Clara: The Academy For Future Science, 1987.

Imbrie, John and Katherine. *Ice Ages: Solving the Mystery.* Enslow Publishers, 1979.

Jasmuheen. *Living on Light.* Burgrain, Germany: Koha Verlag, 1998.

Joseph, Lawrence E. *Apocalypse 2012: A Scientific Investigation Into Civilization's End.* New York: Morgan Road Books, 2007.

Kaku, Michio. *Hyperspace: A Scientific Odyssey Through Parallel Universes, Time Warps, and the 10th Dimension.* New York: Anchor Books, 1995.

Kenyon, Tom, and Virginia Essene. *The Hathor Material: Messages from an Ascended Civilization.* Santa Clara: Spiritual Education Endeavors, 1996.

King, Godfre Ray. *Unveiled Mysteries.* Dunsmuir, CA: St. Germain Press, 1934.

Kolbert, Elizabeth. *The Sixth Extinction: An Unnatural History*. London: Bloomsbury Publishing, 2014.

Lanza, Robert, and Bob Berman. *Biocentrism: How Life and Consciousness are the Keys to Understanding the True Nature of the Universe*. Dallas: BenBella, 2009.

LaViolette, Paul. *Beyond the Big Bang*. Rochester: Park Street Press, 1995.

Ibid. *Earth Under Fire: Humanity's Survival of the Ice Age*. Rochester: Bear & Company, 2005.

Ibid. *Subquantum Kinetics: A Systems Approach to Physics and Cosmology*. Alexandria: Starlane Publications, 2003.

Losey, Meg Blackburn. *The Children of Now: Crystalline Children, Indigo Children, Star Kids, Angels on Earth, and the Phenomenon of Transitional Children*. Franklin Lakes: Career Press, 2007.

Lovelock, James. *The Ages of Gaia: A Biography of Our Living Earth*. New York: Norton, 1995.

Ibid. *Gaia: A New Look at Life on Earth*. New York: Oxford University Press, 1979.

Ibid. *The Revenge of Gaia: Earth's Climate Crisis and the Fate of Humanity*. New York: Basic Books, 2006.

Lynas, Mark. *Six Degrees: Our Future on a Hotter Planet*. London: Harper Perennial, 2008.

MacLean, Dorothy. *To Hear the Angels Sing: An Odyssey of Co-Creation With the Devic Kingdom*. Herndon: Lindisfarne Books, 1994.

Maharaj, Sri Nisargadatta. *I Am That*. Mumbai: Chetana, 1973.

Marciniak, Barbara. *Earth: Pleiadian Keys to the Living Library*. Rochester: Bear & Company, 1994.

Meece, Alan. *Horoscope for the New Millennium*. Woodbury: Llewellyn, 1997.

Melchizedek, Drunvalo. *Ancient Secret of the Flower of Life: Volume 1*. Flagstaff: Light Technology, 1990.

Ibid. *Ancient Secret of the Flower of Life: Volume 2*. Flagstaff: Light Technology, 2000.

Ibid. *Living in the Heart*. Flagstaff: Light Technology, 2003.

Ibid. *The Serpent of Light: Beyond 2012*. San Francisco: Red Wheel/Weiser, 2008.

Mindell, Arnold. *The Shaman's Body: A New Shamanism for Transforming Health, Relationships, and the Community*. New York: HarperCollins, 1993.

Parker, Geoffrey. *Global Crisis: War, Climate Change and Catastrophe in the Seventeenth Century*. The Population Council, 2014.

Perkins, John. *Confessions of an Economic Hit Man*. San Francisco: Berrett-Koehler Publishers, 2004.

Ibid. *Hoodwinked: An Economic Hitman Reveals Why the Global Economy Imploded - and How to Fix It*. Crown Business: 2011.

Ibid. *New Confessions of an Economic Hit Man*. New York: McGraw Hill, 2016.

Phillips, Graham. *Act of God*. London: Pan Books, 1998.

Pierce, Lawrence. *A New Little Ice Age Has Started: How to Survive and Prosper during the Next 50 Difficult Years*. Create Space, 2015.

Rachele, Sal. *Earth Changes and 2012: Messages from the Founders*. Wentworth: Living Awareness Productions, 2008.

Rain, Mary Summer. *Phoenix Rising: No Eyes' Vision of the Changes to Come*. Charlottesville: Hampton Roads, 1987.

Ray, Paul H., & Sherry Ruth Anderson. *The Cultural Creatives: How 50 Million People Are Changing the World*. New York: Three Rivers Press, 2001.

Roads, Michael J. *Journey Into Nature: A Spiritual Adventure*. Tiburon: H.J. Kramer, 1990.

Ibid. *Journey Into Oneness: A Spiritual Odyssey*. Tiburon: H.J. Kramer, 1994.

Ibid *Talking with Nature: Sharing the Energies and Spirit of Trees, Plants, Birds, and Earth*. Tiburon: H.J. Kramer, 1987.

Robbins, Dianne. *TELOS: The Call Goes Out from the Hollow Earth and the Underground Cities.* Mt. Shasta: Mount Shasta Light Publishing, 2000.

Rother, Steve, and the Group. *Re-member: A Handbook for Human Evolution.* Poway: Lightworker, 2000.

Ibid. *Welcome Home: The New Planet Earth.* Poway: Lightworker, 2002.

Russell, Peter. *From Science to God.* Novato: New World Library, 2003.

Ibid. *The Global Brain Awakens: Our Next Evolutionary Leap.* Palo Alto: Element Books, 1995.

Ibid. *A White Hole in Time: Our Future Evolution and the Meaning of Now.* London: HarperCollins, 1992.

Sams, Gregory. *Sun of God: Discover the Self-organizing Consciousness that Underlies Everything.* San Francisco: Wieser Books, 2009.

Sangen, Yvonne. *Biophotons: Source of Energy and Life Light.* Utrecht: Ankh-Hermes, 2010.

Santillana, Giorgio, *Hamlet's Mill: Investigating the Origins of Human Knowledge and Its Transmission through Myth.* David Godine Inc, 1977.

Satprem. *Sri Aurobindo or the Adventure of Consciousness.* Mysore: Mira Aditi, 2003.

Ibid. *The Mind of the Cells or Willed Mutation of Our Species.* Mysore: Mira Aditi, 2002.

Scallion, Gordon Michael. *Notes From the Cosmos: A Futurist's Insights Into the World of Dream Prophecy and Intuition.* Chesterfield: Matrix Institute, 1997.

Scott, Donald. *The Electric Sky: A Challenge to the Myths of Modern Astronomy.* Mikamar Publishing, 2006.

Sheldrake, Rupert. *A New Science of Life: The Hypothesis of Morphic Resonance.* New York: J.P. Tarcher, 1982.

Sitchin, Zecharia. *Genesis Revisited: Is Modern Science Catching Up With Ancient Knowledge?* Rochester: Bear & Company, 2002.

Bibliography

Snow, Chet. *Mass Dreams of the Future*. New York: McGraw Hill, 1989.

Spangler, David. *Revelation: The Birth of a New Age*. Everett: Lorian Press, 1970.

Ibid. *Towards a Planetary Vision*. Findhorn: Findhorn Publications, 1977.

Spaulding, Baird T. *Life and Teaching of the Masters of the Far East*. *6 vols.* Marina del Ray: DeVorss, 1986.

Sri Aurobindo. *The Future Evolution of Man: The Divine Life Upon Earth*. Pondicherry: Sri Aurobindo Ashram Trust, 1963.

Ibid. *The Life Divine*. Pondicherry: Sri Aurobindo Ashram Trust, 1955.

Ibid. *Savitri: A Legend and a Symbol*. Pondicherry: Sri Aurobindo Ashram Trust, 1950.

Steyn, Mark (compiler). *A Disgrace to the Profession: The World's Scientists On Michael E Mann, His Hockeystick, and their Damage to Science*. Stockade Books, 2015.

Stray, Geoff. *Beyond 2012: Catastrophe or Ecstasy – A Complete Guide to End-of-Time Predictions*. East Sussex: Vital Signs Publishing, 2005.

Sultan, Gary and Douglas Vogt. *God's Day of Judgment: The Real Cause of Global Warming*. Bellevue: Vector Associates, 2007.

Ibid. *Reality Revealed: The Theory of Multidimensional Reality*. Bellevue: Vector Associates, 1978.

Swartz, James. *How to Attain Enlightenment: The Vision of Nonduality*. Sentient Publications, 2009.

Talbot, Michael. *Mysticism and the New Physics*. New York: Bantam Books, 1981.

Teilhard de Chardin, Pierre. *The Future of Man*. New York: Image Books, 2004.

Thornhill, Wallace and David Talbot, *The Electric Universe: A New View of Earth, the Sun and the Heavens*. Mikamar Publishing, 2007.

Timms, Moira. *Beyond Prophecies and Predictions: Everyone's Guide to the Coming Changes.* New York: Ballantine Books, 1996.

Vahrenholt, Fritz and Sebastian Luning. *The Neglected Sun: Why the Sun Precludes Climate Catastrophe.* Arlington Heights: Heartland Institute, 2015.

van Vrekhem, Georges. *Beyond Man: The Life and Work of Sri Aurobindo and the Mother.* New Delhi: Rupa & Co, 2007.

Ibid. *Patterns of the Present: From the Perspective of Sri Aurobindo and the Mother.* New Delhi: Rupa & Co, 2002.

Velikovsky, Immanuel. *Ages in Chaos: Volume 1 – From the Exodus to King Akhnaton.* New York: Doubleday, 1952.

Ibid. *Earth in Upheaval.* New York: Doubleday, 1955.

Ibid. *Worlds in Collision.* New York: Doubleday, 1950.

Watson, Lyall. *Supernature: A Natural History of the Supernatural.* London: Hodder & Stoughton, 1973.

Ward, Suzanne. Matthew, *Tell Me About Heaven: A Firsthand Description of the Afterlife.* Camas: Matthew Books, 2002.

Ibid. *Revelations for a New Era: Keys to Restoring Paradise on Earth.* Camas: Matthew Books, 2001.

West, John Anthony. *Serpent in the Sky: The High Wisdom of Ancient Egypt.* Wheaton: Quest Books, 1993.

White Wolf, Helen. *The Ark of Consciousness: 2012 and the Seeding of the New Civilization.* Auckland: Inspired Earth Publishing, 2009.

Weidner, Jay, and Vincent Bridges. *The Mysteries of the Great Cross of Hendaye: Alchemy and the End of Time.* Rochester: Destiny Books, 1999.

Wilcock, David. *The Source Field Investigations: The Hidden Science and Lost Civilizations Behind the 2012 Prophecies,* Dutton, 2012.

Williamson, Marianne. *Healing the Soul of America: Reclaiming Our Voices as Spiritual Citizens.* New York: Simon & Schuster, 2000.

Bibliography

Ibid. *Return to Love: Reflections on the Principles of A Course in Miracles*. New York: HarperCollins, 1992.

Wilson, Colin. *Atlantis and the Kingdom of the Neanderthals*. Rochester: Bear & Company, 2006.

Windrider, Kiara. *Doorway to Eternity: A Guide to Planetary Ascension*. Mt. Shasta: Heaven on Earth Project, 2001.

Ibid. *Ilahinoor: Awakening the Divine Human*. Boulder: Divine Arts, 2012.

Ibid. *Year Zero: Time of the Great Shift*. Boulder: Divine Arts, 2011.

Wright, MacHaelle Small. *Behaving as if the God in All Life Mattered*. Warrenton: Perelandra, 1983.

Yarbrough, Scott. *The New Ice Age: The Truth about Climate Change*. 2015

Yogananda, Paramahansa. *Autobiography of a Yogi*. Los Angeles: Self-Realization Fellowship, 1946.

Yukteswar, Sri. *The Holy Science*. Los Angeles: Self-Realization Fellowship, 1990.

Zukav, Gary. *The Dancing Wu Li Masters: An Overview of the New Physics*. New York: Bantam, 1980.

Ibid. *The Seat of the Soul*. New York: Simon & Schuster, 1989.

SUGGESTED RESOURCES

Websites

2012.com.au — a 2012 Unlimited library

2012timeforchange.com — from conscious evolution to practical solutions

Abruptearthchanges.com – comets and destruction in the 14th Century during the last Little Ice Age in Europe

Abundanthope.net — Urantia material

Adapt2030 – current information on the upcoming Little Ice Age

Ascension2012.com — evolution of consciousness

BarbaraMarxHubbard.com — Foundation for Conscious Evolution

BillHerbst.com — good articles on planetary astrology

Castaneda.com — presenting the foundations of Tensegrity, inspired by Carlos Castaneda

Collective-evolution.com – Heart math and other videos

Commondreams.org — network for co-creating the new world

Crawford2000.co.uk — earth changes news

Cropcircleconnector.com — crop circles newsletter

Culturalcreatives.org — Paul Ray and Sherry Anderson's website

Diagnosis2012.co.uk — Geoff Stray's website

Disclosureproject.org — Steven Greer's website

Divinecosmos.com — David Wilcock's website

Evolutionaryleaps.com — Robert Felix's website

Experiencefestival.com — experience of global oneness

Earthchangesmedia.com — Mitch Battros' Earth science news

Enlightened-spirituality.org — spiritual awakening

Etheric.com — Paul LaViolette's website

Freeenergytruth.blogspot.com — alternate and free energy website

GlobalOnenessProject.com — a film library representing interconnectivity in today's complex world

Gnosticmedia.com — esoteric and shamanic news and interviews

Grahamhancock.com — Graham Hancock's website

Greggbraden.com — Gregg Braden's website

Groups.yahoo.com/group/earthchange-bulletins — Michael Mandeville on Earth changes

Globalfamily.net — education and support for personal and planetary transformation

Greatdreams.com — dreams of the great Earth changes

Halexandria.org — collection of articles on ancient Egypt

Handpen.com — Paul Winter's website

Hollowplanets.com — a feasibility study of hollow worlds

Iceagenow.com — research on imminent ice age

Igc.org — Institute for Global Communications, connecting people who are changing the world

Integralyoga-auroville.com — about Sri Aurobindo and the Mother

Itsrainmakingtime.com — Kim Greenhouse on breakthroughs, conversations, discoveries

Jayweidner.com — Jay Weidner's website

Jeanhouston.org — Jean Houston's website

Keysofenoch.com — J.J. Hurtak's website

Kiarawindrider.net — Kiara Windrider's website

Kryon.com — Lee Carroll's website

Leadingedgenews.com — alternate news

Lightparty.com — health, peace, and freedom for all

Lightworker.com — Steve Rother's website

Luminousself.com — Barry Martin and Karen Anderson's website

Magneticreversal.org – Ben Davidson and Mitch Battros website

Matthewbooks.com — Suzanne Ward's website

Mindsing.org — in pursuit of the consciousness singularity

Mkaku.org — Michio Kaku's website

Monbiot.com — George Monbiot's website

Nealadams.com — research and animation of expanding Earth theory

Newenergymovement.org — alternative energies

Niburu.nl — alternative news website in Dutch and English

Operationterra.com — Lyara's website on Earth changes, ETs, the end times, and the journey to the New Earth

Ourhollowearth.com — explorations of the hollow Earth theory, a corollary to the expanding Earth theory

Peterussell.com — Peter Russell's website

Projectcamelotportal.com — Fascinating interviews

Raisingloveconsciousness.com — Kashonia Carnegie's blog

Rense.com — Jeff Rense's alternative news website

Selfempowermentacademy.com — Jasmuheen's website

Sheldrake.org — Rupert Sheldrake's website

Sitchin.com — Zechariah Sitchin's website

Soulutions.co.uk — Soleira and Santara Green's evolutionary website

Spiritofmaat.com — Drunvalo Melchizedek's website

Swirlednews.com — crop circles newsletter

Thevenusproject.com — Jacque Fresco's website on planetary redesign

Thezeitgeistmovement.com — Peter Joseph's website advocating a resource-based economy

Thomhartmann.com — Thom Hartmann's website

Timeofglobalshift.com — a holistic, spiritual view of the present time on Earth, and the role we each can play to support the current global transformation

Tomkenyon.com — planetary messages from the Hathors

Trackerschool.com — Tom Brown Jr.'s website

Webdirectory.com — Earth's biggest environmental search engine

Wingmakers.com — Time capsule from a future version of humankind

Wisdomsplendour.org — Sri Aurobindo, the Mother, and Integral Yoga

Wisdomuniversity.org — university founded by Barbara Marx Hubbard

Worldpuja.org — live internet webcasts for peace

Movies

Avatar (glimpse of Gaia and the interconnectedness of life)

Brother Sun, Sister Moon (St. Francis of Assisi)

Dances with Wolves (clash of cultures)

Deep Impact (meteorite impact scenario)

Dragonfly (power of dreams)

Harry Potter series (latent magical abilities)

Illusion (akashic records)

Phenomenon (awakening extrasensory gifts)

Powder (supramental human)

Sliding Doors (multiple timelines)

The Big Blue (human-dolphin interactions)

The Core (Core warming)

The Day After Tomorrow (Ice Age scenario)

The Lake House (parallel timelines)

The Last Avatar (awakening the lightbody)

Whale Rider (human-whale interactions)

Documentaries

Awake in the Dream (Catharina Roland — Experience the Shift)

Earth Under Fire (Paul LaViolette's work)

The Corporation (politics behind the scenes)

The Secrets of Alchemy (Jay Weidner)

What the Bleep Do We Know? (new science)

What the Bleep: Down the Rabbit Hole (more new science)

Why We Fight (military-industrial complex)

Zeitgeist: The Movie, Zeitgeist: Addendum, and Zeitgeist: Moving Forward (directed by Peter Joseph — people's movement for socio-economic change)

GLOSSARY

Animal Human — The first of four stages of human evolution, according to Sri Aurobindo. At this stage our primary goal is survival of the self and survival of the species. It is based on instinctive reactions and behaviors. Taken to an extreme, this results in the experience of separateness and perpetual conflict.

Ascended Master — One who has succeeded in unifying the physical body with the light body.

Ascension — This term has both personal and planetary components. It refers to the process of raising the frequencies of the physical body so as to merge with its higher-dimensional counterpart, the light body, thus transcending the ordinary limitations of a third-dimensional body. The term is also used to refer to our translation from third-dimensional consciousness to fourth, fifth, or sixth-dimensional consciousness (See Shift of the Ages).

Atlantis — A continent hosting an advanced race of humans believed to have sunk 12,000 years ago. The sinking of Atlantis could be related to both a magnetic reversal between 11,800 and 12,400 years ago and to a corresponding pole shift.

Axiatonal Lines — In The Keys of Enoch, J.J. Hurtak refers to these higher-dimensional flows of energy which serve as galactic extensions of the meridian system in the body. When fully open, it enables one to heal the body, regenerate limbs, and even raise the dead.

Base Harmonic Frequency — The vibrational resonance inherent in matter. Alexey Dmitriev postulates that the increase in cosmic dust within our Solar System would raise the base harmonic frequency of matter. This closely correlates with Sri Aurobindo's references to an imminent transformation of our bodies into true matter.

Blue Star — There is a Hopi prophecy that the Fifth World would be birthed following the arrival of the Blue Star. LaViolette believes that this phenomenon is related to periodic explosions in the galactic core which emit massive quantities of cosmic energies. This pulse of cosmic energy becomes visible to us through layers of interstellar dust as a brilliant bluish-white star shortly before its impact reaches our Solar System. LaViolette speculates that the next pulse of this galactic superwave is imminent.

Central Sun — A theosophical term for both the physical and interdimensional aspects of the consciousness embodying the center of our galaxy.

Cosmic Dust — Refers to interstellar debris in outer space, including fragments of exploded stars and planets. Dmitriev has been researching the effects of an increase of cosmic dust recently encountered within our Solar System.

Crop Circles — Increasingly complex geometrical patterns that have been mysteriously created in wheat fields all over the world, but most famously in Wiltshire County, UK. There are many theories as to how they are created and what they mean, including the plasma theory — under certain conditions, noospheric information can be physically out pictured as geometrical patterns on the surface of the Earth and as new genetic patterns within our DNA.

Crustal Plate Displacement Theory — Charles Hapgood proposes that under certain conditions the crust of the Earth is capable of slipping, causing a dramatic shift in the location of oceans and continents on the Earth's surface. In contrast to plate tectonics, crustal displacement indicates that the entire crust of the Earth is capable of disengaging from the lower crust and mantle and shifting together (like the loose skin on a tangerine). The last time this happened was 12,000 years ago and contributed to the displacement and sinking of Atlantis.

Crystal Children — A next generation of children, after the Indigo Children, who bring powerful gifts and abilities to help in Earth's awakening. They will set the stage for a new species of supramental beings on Earth.

Dimensions — A measure of vibrational frequency, or density. In our involutionary journey, we created and descended through various dimensional levels eventually reaching the third dimension. Our "fall" into this linear dimension of space and time provided the experience of extreme vibrational density and a duality between good and evil. In our evolutionary journey back to Source, we are gradually raising vibrational frequency, moving past the lessons of duality (karma), reconnecting with our souls, and moving into ascension consciousness. The Shift of the Ages is associated with rising up through the dimensions.

Third Dimension — Characterized by separation from soul. Duality. The experience of time is linear. Our physical and planetary bodies are governed by the law of increasing chaos, or entropy.

Fourth Dimension — A transitional dimension in which physical density is loosened and time is increasingly experienced as synchronistic.

Fifth Dimension — Characterized by a more unified consciousness within and without, reflected in an Age of Light. Many parallel realities exist here and inter-dimensional travel becomes possible. It is the realm of Sri Aurobindo's overmental consciousness.

Sixth Dimension — Ascension mastery becomes possible on both personal and planetary levels. It is the realm of Sri Aurobindo's supramental consciousness.

Divine Human — The third stage of our current human evolution according to Sri Aurobindo. A stage where our soul consciousness can be fully embodied in our current human forms.

Galactic Center — A black hole, referred to as the Central Sun in theosophical literature, located at 26 degrees of Sagittarius. Personified as Hunab Ku. The origin of LaViolette's galactic superwaves.

Galactic Superwave Theory — According to Dr. Paul LaViolette, there is a massive emanation of energy pulsing out from our galactic core about every 12,000 years. Each pulse lasts several hundred to several thousand years. This is due to reach us again shortly. May be compared with the Blue Star Kachina of Hopi prophecy, which would herald our journey into the Fifth World.

Gamma Ray Bursts (GRBs) — More powerful than any supernova, these extra-galactic fireballs could be the result of mutually annihilating neutron stars colliding at the centers of spiral galaxies.

Global Brain — The morphogenetic field of a unified planet, this term is used by Peter Russell to represent the next level of species and planetary evolution following a quantum leap in consciousness in which our primary identity would be global rather than individual. The Internet is a technological representation of this global brain.

Hollow Earth Theory — This controversial theory holds that Earth, along with the other planets, has a hollow core and Inner Sun resulting from its centrifugal motion as gases cooled and lava began to solidify. Civilizations exist within this inner Earth, which can be accessed through openings at the North and South Poles.

Human Human — The second stage of our current human evolution, according to Sri Aurobindo. This is a transitional stage where our souls begin to embody third-dimensional matter.

Hunab Ku — Mayan-attributed term referring to the center of everything, whether reflected in the center of the universe as the Source and Creator of all there is, or within the heart of each atom. Usually referring to a galactic consciousness physically represented by the Galactic Center.

Indigo Children — Also known variably as the Rainbow Children or Crystal Children, a transitional species of humans who are being born in our midst today. They are the forerunners of an Age of Peace.

Light body — A vehicle for our indwelling souls more fluid and responsive to our multidimensional impulses than is the current state of our physical body. According to Sri Aurobindo and other spiritual teachers, we are moving towards a time when the light body will fully merge with the physical body.

Mind — Most of us equate this with the rational mind, which is simply one layer of perception in a multidimensional universe. Beyond this, according to Sri Aurobindo, are many subtler layers of mind, capable of experiencing reality in very different ways. This includes the Higher Mind, the Illumined

Mind, the Intuitive Mind, the Overmind, and the Supermind. We become capable of experiencing increasingly subtle dimensions of reality as we get in touch with the corresponding layers of Mind. There is also the Cellular Mind, or the intelligence held within the body itself.

Morphogenetic Fields — According to Rupert Sheldrake, these form-generating fields provide the basis for our memories of the past and our possibilities for the future. The Akashic Records, said to contain a record of everything that has ever transpired, are a morphogenetic field inter-penetrating all matter, but especially accessible within the noosphere. The hundredth monkey theory is an example of how these fields can serve an evolutionary purpose.

Noosphere — A term developed by French paleontologist and Jesuit priest Pierre Teilhard de Chardin. Describes a morphogenetic field of evolutionary intelligence physically held within the ionosphere of the Earth.

Photon Belt — A common new age metaphysical theory that refers to our encounter with a band of high frequency photons about every 13,000 years. The potential effects of such an encounter would be either cataclysmic destruction or spiritual rebirth, or both. Although first used in the context of science fiction, the term does not have much scientific backing. Some elements of this theory can be seen reflected in various cosmic phenomena.

Pole Shift — Refers to either a geographical pole shift, a magnetic reversal, or both simultaneously. There is evidence that a magnetic reversal occurs nearly every 12,000 years. Geographical pole shifts could follow magnetic reversals as the Earth realigns herself. Shifts could occur independently, as when asteroids have collided with Earth in the past. Either of these types of pole shifts could mean cataclysmic changes on Earth's surface, the extent of which would be determined by the current density of consciousness on the planet. We are due for a pole shift soon according to many sources.

Precessional Year — The seeming movement of the constellations in the Sky relative to us, based on the tilt of the

Earth's axis. Astrological world ages are computed according to this 26,000-year cycle.

Schumann Resonance — Refers to the base frequency of the Earth, measured at 7.83 hertz, or cycles per second. Gregg Braden, drawing on research by Russian and Swedish scientists, assesses that this frequency, stable for thousands of years, is now rising towards 13 hertz — the frequency of an awakened Earth.

Shift of the Ages — Refers to the collective translation of our planet into an Age of Light marked by a higher-vibrational fourth dimension or fifth-dimension frequency. There are many layers of factual significance. The term may be associated with one or all of the following components:

Moving from kali yuga to satya yuga.

Moving from the Age of Pisces, comprising 2,160 years of precession to the Aquarian Age.

Ending of a Long Count from 3114 BC to AD 2012 in the Mayan calendar, and entering the Sixth Sun or Solar Age.

Moving from the Fourth World to the Fifth World of the Hopis, following the appearance of the Blue Star.

Shifting of the geographical poles.

Shifting or reversal of the magnetic poles.

Zero Point, as defined by Gregg Braden — the simultaneous raising of Earth's base frequency and the collapsing of her magnetic fields.

Vibrational shift from the third dimension to higher dimensions.

Supramental — A state of unified harmony that exists beyond the mental ego of our ordinary human consciousness.

Supramental Human — A future race on Earth, some 300 years hence, in which an expanded soul consciousness would fully incarnate within bodies of true matter, according to Sri Aurobindo and the Mother. This is the fourth and final stage of our current human evolution.

Supramental Descent — The term used by Sri Aurobindo and the Mother to indicate the quantum awakening in Earth's future — higher spiritual worlds are unified within cellular consciousness, human bodies, and all matter on Earth signaling the beginning of a new evolution and corresponding with fifth and sixth-dimensional consciousness.

True Matter — A term used by Sri Aurobindo and the Mother pertaining to a subtle frequency of etheric-physical matter. This more elastic dimension of matter would be suitable for the manifestation of multi-dimensional consciousness, culminating in the creation of the supramental human.

Year Zero — The term used by Don Alejandro to represent a time of planetary shift soon expected. The Mayas prefer to use this term rather than to fix the Mayan end-date into Gregorian time.

Yugas — The Hindu conception of world ages. The cycle spans a diminishing from the satya yuga, an age of light, through the treta, dwapara, and finally the kali yuga, an age of darkness. Sri Yukteswar claimed that each complete yugic cycle lasts 24,000 years, during which we twice traverse the succession of yugas. A shift from a kali yuga to a satya yuga is soon anticipated. We are currently undergoing a "purification," a necessary aspect of this Shift.

Zero Point — This term has several layers of meaning:

Synchronization of various cycles of time (Windrider).

A shift associated with the raising of Earth's base frequency and the collapsing of her magnetic fields (Braden).

A still point of infinite creative potential revealed in the moment of magnetic reversal.

A reset button for planetary or stellar consciousness (Kiara and Barry).

A form of free energy technology.

INDEX

301

ABOUT KIARA WINDRIDER

Kiara Windrider spent much of his early life traveling and practicing various spiritual traditions in India. A lifelong interest in environmental awareness, peace making, and social justice led to a dual degree in Peace studies and International Development through Bethel College, North Newton, Kansas.

Later, he completed a graduate program in Transpersonal Counseling Psychology through JFK University in Orinda, California, and worked for many years at an alternative psychiatric center called Pocket Ranch Institute, which specialized in healing emotional trauma and facilitating spiritual emergence. He received a psychotherapy license (MFT) from the State of California in 1998. He has also trained in various forms of bodywork, breathwork, hypnotherapy, and shamanic healing.

Kiara is an avid science researcher, exploring connections between galactic cycles, climate change, ancient history, quantum physics, human behavior and spiritual awakening. As an outcome of this extensive research he has come to the firm conviction that we stand collectively at the brink of a quantum evolutionary leap beyond our wildest dreams.

He is currently focused on planetary healing using a system of anchoring divine light known as Ilahinoor. He also teaches Inka shamanic practices, into which he was initiated by Juan Nunez del Prado and Ivan Nunez del Prado. Kiara has worked with Egyptian, Huna and Sufi traditions, and is also rooted in Integral Yoga and the Advaita traditions of India.

He offers workshops and retreats worldwide for awakening to our infinite potential. His greatest wish is to live fully in the wonder of each moment, and to help awaken this beautiful planet to its luminous destiny.

Kiara is the author of Doorway to Eternity: A Guide to Planetary Ascension, Deeksha: Fire from Heaven, Journey into Forever: Surfing 2012 and Beyond, Year Zero: Time of the Great Shift, Ilahinoor: Awakening the Divine Human, and Issa: Son of the Sun.

Please check out his website, Kiarawindrider.net. Suggestions and comments are always welcome, and may be addressed to kiarawindrider@gmail.com.

Kiara was born on March 6, 1959 at 2:06 a.m. in Bombay, India.

Made in the USA
Middletown, DE
11 November 2022

14753123R00182